AF735941

VOLUME FORTY FOUR

THE ENZYMES

Mesoporous Silica-based Nanomaterials and Biomedical Applications, Part B

VOLUME FORTY FOUR

THE ENZYMES

Mesoporous Silica-based Nanomaterials and Biomedical Applications, Part B

Edited by

FUYUHIKO TAMANOI
Institute for Integrated Cell-Material Sciences, Institute for Advanced Study, Kyoto University, Kyoto, Japan;
Department of Microbiology, Immunology and Molecular Genetics, University of California, Los Angeles, CA, United States

ELSEVIER
ACADEMIC PRESS
An imprint of Elsevier

Academic Press is an imprint of Elsevier
50 Hampshire Street, 5th Floor, Cambridge, MA 02139, United States
525 B Street, Suite 1650, San Diego, CA 92101, United States
The Boulevard, Langford Lane, Kidlington, Oxford OX5 1GB, United Kingdom
125 London Wall, London, EC2Y 5AS, United Kingdom

First edition 2018

Copyright © 2018 Elsevier Inc. All rights reserved.

No part of this publication may be reproduced or transmitted in any form or by any means, electronic or mechanical, including photocopying, recording, or any information storage and retrieval system, without permission in writing from the publisher. Details on how to seek permission, further information about the Publisher's permissions policies and our arrangements with organizations such as the Copyright Clearance Center and the Copyright Licensing Agency, can be found at our website: www.elsevier.com/permissions.

This book and the individual contributions contained in it are protected under copyright by the Publisher (other than as may be noted herein).

Notices

Knowledge and best practice in this field are constantly changing. As new research and experience broaden our understanding, changes in research methods, professional practices, or medical treatment may become necessary.

Practitioners and researchers must always rely on their own experience and knowledge in evaluating and using any information, methods, compounds, or experiments described herein. In using such information or methods they should be mindful of their own safety and the safety of others, including parties for whom they have a professional responsibility.

To the fullest extent of the law, neither the Publisher nor the authors, contributors, or editors, assume any liability for any injury and/or damage to persons or property as a matter of products liability, negligence or otherwise, or from any use or operation of any methods, products, instructions, or ideas contained in the material herein.

ISBN: 978-0-12-815111-2
ISSN: 1874-6047

For information on all Academic Press publications
visit our website at https://www.elsevier.com/books-and-journals

ELSEVIER
Book Aid International
Working together to grow libraries in developing countries
www.elsevier.com • www.bookaid.org

Publisher: Zoe Kruze
Acquisition Editor: Sam Mahfoudh
Editorial Project Manager: Shellie Bryant
Production Project Manager: Abdulla Sait
Cover Designer: Victoria Pearson

Typeset by SPi Global, India

CONTENTS

CONTRIBUTORS

Montserrat Colilla
Department of Chemistry in Pharmaceutical Sciences, School of Pharmacy, Universidad Complutense de Madrid and Networking Research Center on Bioengineering, Biomaterials and Nanomedicine (CIBER-BBN), Madrid, Spain

Tan Le Hoang Doan
Institute for Integrated Cell-Material Sciences, Institute for Advanced Study, Kyoto University, Kyoto, Japan; Center for Innovative Materials and Architectures (INOMAR); Faculty of Chemistry, University of Science, Vietnam National University—Ho Chi Minh City, Ho Chi Minh City, Vietnam

Carlotta A. Glackin
Developmental and Stem Cell Biology, City of Hope Medical Center, Duarte, CA, United States

Shinji Inagaki
Toyota Central R&D Labs., Inc., Nagakute, Japan

Soontaree (Grace) Intasa-ard
School of Molecular Science and Engineering, Vidyasirimedhi Institute of Science and Technology (VISTEC), Wangchan, Rayong, Thailand

Isabel Izquierdo-Barba
Department of Chemistry in Pharmaceutical Sciences, School of Pharmacy, Universidad Complutense de Madrid and Networking Research Center on Bioengineering, Biomaterials and Nanomedicine (CIBER-BBN), Madrid, Spain

Aoi Komatsu
Institute for Integrated Cell-Material Sciences, Institute for Advanced Study, Kyoto University, Kyoto, Japan

Kazuyuki Kuroda
Faculty of Science and Engineering; Kagami Memorial Research Institute for Materials Science and Technology, Waseda University, Tokyo, Japan

Ngoc Xuan Dat Mai
Institute for Integrated Cell-Material Sciences, Institute for Advanced Study, Kyoto University, Kyoto, Japan; Center for Innovative Materials and Architectures (INOMAR), University of Science, Vietnam National University—Ho Chi Minh City, Ho Chi Minh City, Vietnam

Kotaro Matsumoto
Institute for Integrated Cell-Material Sciences, Institute for Advanced Study, Kyoto University, Kyoto, Japan

Keigo Nakai
Institute for Integrated Cell-Material Sciences, Institute for Advanced Study, Kyoto University, Kyoto, Japan

Michihiro Nakamura
Department of Organ Anatomy & Nanomedicine, Yamaguchi University Graduate School of Medicine, Ube, Japan

Makoto Ogawa
School of Energy Science and Engineering, Vidyasirimedhi Institute of Science and Technology (VISTEC), Wangchan, Rayong, Thailand

Fuyuhiko Tamanoi
Institute for Integrated Cell-Material Sciences, Institute for Advanced Study, Kyoto University, Kyoto, Japan; Department of Microbiology, Immunology and Molecular Genetics, University of California, Los Angeles, CA, United States

María Vallet-Regí
Department of Chemistry in Pharmaceutical Sciences, School of Pharmacy, Universidad Complutense de Madrid and Networking Research Center on Bioengineering, Biomaterials and Nanomedicine (CIBER-BBN), Madrid, Spain

Eisuke Yamamoto
Institute of Materials and Systems for Sustainability, Nagoya University, Chikusa-ku, Japan

PREFACE

This volume is Part B of a two-volume project entitled, "Mesoporous Silica-Based Nanomaterials and Biomedical Applications." In the previous volume (Part A, Volume 43), we started with an overview of research on mesoporous silica nanoparticles (MSN), discussed structure characterization, and emphasized MSN's feature that enables designing controlled drug delivery capability. In this volume, we discuss controlled synthesis of MSN. Preparation of periodic mesoporous organosilica (PMO) is also discussed. We then discuss application of MSN for bone infection treatment as well as for cancer therapy. As for cancer therapy application, we describe studies using mouse models as well as a strategy to shut down gene expression by delivering siRNA by MSN. The theme of biodistribution of MSN is revisited in this volume by the review of tumor accumulation of MSN. The issue of biodegradability of MSN is revisited by the discussion of the recent work on characterizing anticancer drug delivery capability of biodegradable PMO in the chicken egg tumor model. Finally, we discuss other materials including layered silicate and organosilica nanoparticles.

We would like to thank all the authors for preparing their chapters despite their busy schedule. Without their timely preparation, we could not have put together this volume. We would also like to thank editorial help by Shellie Bryant of Elsevier.

Fuyuhiko Tamanoi
September 1, 2018
Kyoto and Los Angeles

CHAPTER ONE

Preparation and Controllability of Mesoporous Silica Nanoparticles

Eisuke Yamamoto*, Kazuyuki Kuroda†,‡,1
*Institute of Materials and Systems for Sustainability, Nagoya University, Chikusa-ku, Japan
†Faculty of Science and Engineering, Waseda University, Tokyo, Japan
‡Kagami Memorial Research Institute for Materials Science and Technology, Waseda University, Tokyo, Japan
[1]Corresponding author e-mail address: kuroda@waseda.jp

Contents

Abstract

Mesoporous silica nanoparticles (MSNs) have attracted great attention because of their high surface area, large pore volume, transparency, biocompatibility, and high cell-uptake efficiency. Such unique properties allow the use of these materials such as catalysts, drug carriers, and optical materials. This chapter introduces the preparation of MSNs and some recent developments in the preparation of MSNs with precisely controlled particle size, pore size, functionality, and morphology.

1. INTRODUCTION

Silica-based materials have been utilized in diverse applications such as catalysts, fillers, adsorbents, etc. Their morphology and structure strongly affect the functions. In the late 1980s, one of the authors and his group studied various interactions of silicates with organic compounds [1]. Those studies accidentally led to the discovery of mesoporous silica from the layered silicate, kanemite ($NaHSi_2O_5 \cdot 3H_2O$). Alkyltrimethylammonium–kanemite

The Enzymes, Volume 44
ISSN 1874-6047
https://doi.org/10.1016/bs.enz.2018.09.001
© 2018 Elsevier Inc.
All rights reserved.

complexes formed three-dimensional silica networks and were calcined to yield mesoporous silica. The mesopore size was simply controlled using the chain length of the alkyltrimethylammonium ions used. MCM-type mesoporous silica was reported by Mobil in 1992, and this sparked huge numbers of studies related to mesoporous silica [2]. The control of the morphology and structure of mesoporous silica is one of the important studies for the effective use of these unique materials. In 1997, Unger et al. reported the preparation of spherical mesoporous silica particles using the Stöber method [3]. Since this report, a variety of mesoporous silica particles have been prepared. Furthermore, downsizing mesoporous silica to the nano-size opened up novel applications for it. In particular, mesoporous silica nanoparticles (MSNs) are expected to be useful in biomedical applications because of their unique characteristics, including high cell uptake ability [4–13]. In addition, their functions and properties can be easily controlled by changing the particle size, pore size, morphology, and surface (Fig. 1). At the early stage of the studies on MSNs, Mann et al. prepared MSNs that were several tens of nanometers in size [14], and Cui et al. found that MSNs can be prepared by synthesizing nanoparticles in a system with diluted concentrations of surfactants [15]. In this chapter, we introduce the recent developments in the preparation of MSNs, including our recent results.

2. CONTROL OF MESOSTRUCTURE AND DISPERSIBILITY OF MSNs

Mesostructured materials composed of silica and surfactants are prepared by the hydrolysis and condensation of tetraalkoxysilanes ($Si(OR)_4$, R = Me, Et, etc.) in solutions containing various kinds of surfactants (typically, cationic surfactants such as cetyltrimethylammonium bromide (CTAB) or triblock copolymers such as P123) which act as templates for the mesostructure. Nanoparticles smaller than 100 nm have been prepared

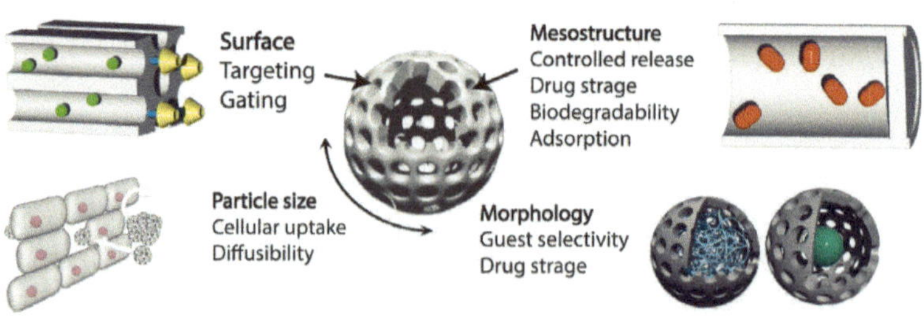

Fig. 1 Relationship between the features of MSNs and their properties.

by appropriately changing the conditions, such as the pH and concentrations of the surfactants and Si species [16–18]. MSNs are obtained by the removal of these surfactants (templates) through calcination or extraction. MSNs have attracted great attention as drug carriers [5,6,8–10,12,13], fillers of polymer composites [19,20], catalyst supports/carriers [21], and photonic crystals [22,23] because of their unique properties, including low toxicity, high surface area, large pore volume, and the functionality due to the presence of surface silanol groups. Recently, MSNs with diverse morphologies and mesostructures (typically, MSNs with cylinder-like pores, ~2–4 nm in size, and wormhole-like pores) have been prepared. The arrangement of the mesopores has also been controlled; typically, they have been hexagonal [24], central-radial [25], and random structures [26]. The morphology of these nanoparticles is almost sphere-like, although rod-like and polyhedron nanoparticles can also be obtained through the control of the mesostructure and arrangement of pores. In addition, a unique morphology, i.e., MSNs with Janus-like structure [27–31], containing two compartments consisting of both cubically and hexagonally arranged mesopores within one particle [32], has been reported. Furthermore, hollow-structured and rattle-type structured MSNs [33] have been prepared for encapsulating a large amount of guest species.

The dispersibility of MSNs is one of the important factors for their applications. Normally, colloidal MSNs have been prepared by dispersing the aggregated nanoparticles using a sonication process. For obtaining MSNs with a high dispersibility, it is essential to induce repulsion forces among the nanoparticles that are stronger than the attractive force (typically, van der Waals interaction) (Fig. 2). The dispersibility of MSNs strongly depends

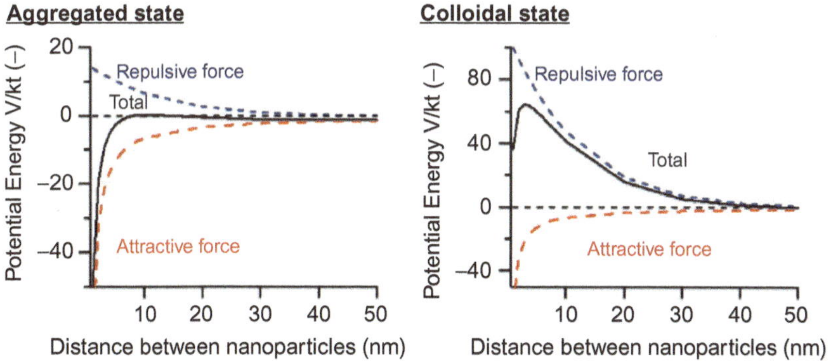

Fig. 2 Differences between the forces among MSNs in the aggregated and colloidal states.

on the template removal process. Although calcination is one of the general methods to remove templates, this process promotes the interparticle condensation of silanol groups among MSNs. In general, the extraction of templates through centrifugation and redispersion has been applied for the preparation of MSNs with a high dispersibility, to avoid irreversible aggregations. However, this centrifugation method is not effective for MSNs smaller than 50 nm in size, because small nanoparticles cannot be redispersed by ultrasonication. A dialysis process is an alternative method for obtaining small MSNs with a high colloidal dispersibility [34]. Our group reported the preparation of colloidal MSNs through the dialysis process against a mixture composed of aq. acetic acid and ethanol. The complete removal of surfactants by repeated dialysis under careful conditions is needed for biomedical applications [35,36]. Bein et al. showed that the template of MSNs can be removed without any aggregation by heating the colloidal mesostructured materials composed of silica and surfactants in a high-boiling-point solution, thereby retaining their dispersibility [37]. Although, basically, these small-sized nanoparticles cannot be redispersed, Haynes et al. succeeded in the redispersion of aggregated small MSNs by functionalizing the surface with trimethylsilyl and polyethylene glycol groups [38]. The design for redispersibility is one of the great issues in using MSNs for practical applications.

3. CONTROL OF PARTICLE SIZE AND PORE SIZE OF MSNs

It is important to precisely control the size of MSNs, for both their applications and an understanding of their properties. In general, the size of MSNs can be varied by appropriately controlling the nucleation and growth by tuning the hydrolysis rate of the alkoxysilanes. This rate can be controlled by simply changing the amount and type of the base catalyst and surfactant, although size control is not so simple, because the variation of these factors also changes the condensation rate of Si species. Our group reported that the type of alkoxy group used (e.g., Si-OMe, Si-OEt, Si-OPr, and Si-OBu) affects the size of the MSNs, and the size becomes larger when alkoxysilanes with longer alkyl chains are used, which may be due to their lower hydrolysis rates (Fig. 3, left) [39]. Furthermore, the addition of alcohol as a cosolvent also leads to a change in the particle size because of the change in the hydrolysis rates.

Although a narrow particle size distribution is essential for applications, the simultaneous control of particle size and distribution is not very easy. A seed-growth method is one of the effective ways to obtain size-controlled

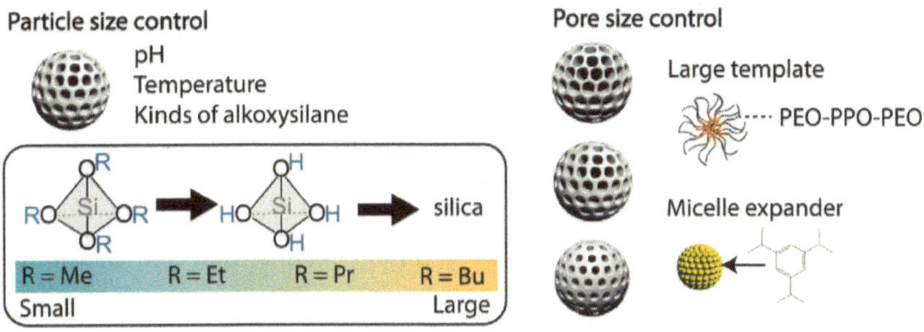

Fig. 3 Control of particle and pore sizes.

MSNs with a narrow size distribution (herein, we define MSNs with a narrow size distribution as monodispersed MSNs). MSNs with narrow size distributions (standard deviation: 5 nm) can be prepared using tetraethoxysilane ($Si(OEt)_4$; TEOS) as seed nanoparticles [40]. The seed nanoparticles were grown using tetrapropoxysilane ($Si(OPr)_4$; TPOS), without any spontaneous nucleation. The obtained MSNs had a high monodispersity and the size of the MSNs could be precisely controlled by changing the amount of TPOS, which allows the formation of colloidal crystals composed of variously sized MSNs.

Furthermore, pore expansion of MSNs is important for the introduction of large-sized guest species, such as DNA [41,42]. There are various methods for expanding the pore size, e.g., the templating method using large-sized templates such as emulsions [43,44] and polystyrene nanoparticles [45]. The addition of a micelle expander is one of the easiest ways to obtain MSNs with large pores (Fig. 3, right) [46,47]. In particular, the addition of trialkylbenzene is effective for the expansion of micelles, and various types of MSNs with large pore sizes have been prepared [48,49]. The procedure of this pore size expansion contains only easy steps, although it is not easy to control both the particle and pore sizes simultaneously. We clarified that the pore size of MSNs can be changed from 4 to 8 nm by changing the amount of triisopropylbenzene (TIPB), while the particle size also changed from 50 to 380 nm [39]. It is effective to use the seed-growth method for controlling both the particle and pore sizes simultaneously. When both TPOS and TIPB were added during the seed-growth period, the particle and pore sizes can be controlled independently by changing the amounts of TPOS and TIPB. The particle size can be controlled from 60 to 100 nm, and the pore size can be controlled from 4 to 10 nm and several nm while retaining their monodispersity [50].

4. FUNCTIONALIZATION OF MSNs

Modification of the interior and exterior of MSNs with organic groups allows the control of the hydrophobicity and dispersibility of MSNs, and various functions such as targeting ability and biocompatibility can be introduced into MSNs. For example, the biocompatibility of MSNs can be increased by surface modification with polyethylene groups. In addition, it has been reported that the effect of reducing a tumor is increased by immobilizing folic acid as a ligand on the surface of particles containing an anticancer agent [51]. Furthermore, the controlled release of a drug by disulfide bond cleavage has been achieved by modifying the surface of the particles with organic groups having disulfide bonds [52].

Organotrialkoxysilanes ($RSi(OR')_3$, $R'=$ Et or Me) are generally used for such organic modifications. There are two modification methods. One is a cocondensation method in which tetraalkoxysilane and organotrialkoxysilane are used for the formation of MSNs with organic modification, and the other is a postgrafting method in which the surfaces of the MSNs are modified after the formation of the MSNs. Both these methods allow the introduction of various organic groups with stable Si–C bonds. In the former method, various functional groups can be easily introduced, and MSNs with a core–shell structure can also be prepared by adding multiple types of organoalkoxysilanes at different timings [53]. Such core–shell structured MSNs with multiple functional groups are particularly useful in DDS applications because this application requires MSNs to exhibit various functions simultaneously [54,55]. However, the cocondensation method has a limitation on the types and amount of introducible functional groups. Furthermore, the addition of organoalkoxysilane affects various parameters, such as the particle size, morphology, and dispersibility, which makes the preparation process of MSNs complex. On the other hand, in the postgrafting method, various kinds of the functional groups can be introduced while retaining the particle size, pore size, and morphology. Unfortunately, the self-condensation of silylating agents in the presence of water in colloidal solutions cannot be avoided easily. Furthermore, the surface modification of organic groups tends to cause the aggregation of particles in water. Hoshikawa et al. dispersed mesostructured nanoparticles with a mixed solution of 2-propanol, hexamethyldisiloxane ($Me_3Si\text{-}O\text{-}SiMe_3$), and hydrochloric acid, for the modification with trimethylsilyl groups (called the Lenz method) [56]. Organically modified

MSNs dispersed in alcohol can be obtained by the simple treatment. Such MSNs are useful as nano fillers with high porosity and can be used for the preparation of an antireflection film by uniform dispersion in a matrix.

5. MORPHOLOGICAL VARIATIONS OF MSNs USING ORGANOSILANE

The morphology of MSNs should be appropriately controlled, depending on the applications. For example, the introduction of hollow structures leads to an increase in the encapsulation amount of guest species. Conventionally, hollow-structured MSNs have been prepared by either a soft template method using block copolymer micelles or a hard template method using nanoparticles [57]. For example, Ikeno et al. prepared hollow-structured MSNs using iron oxide (α-Fe_2O_3) nanoparticles as a hard template and removed the iron oxide nanoparticles by acid etching [58]. Although the hard template method is a slightly complex method, the preparation of hollow organosiloxane nanoparticles using silica particles as a template is a simple, low cost, and highly controllable method. An organosiloxane shell can be formed on silica nanoparticles, using bridged organoalkoxysilane ($(EtO)_3Si$–R–$Si(OEt)_3$, R = –C_2H_4– or –C_6H_4–). After the formation of core–shell structured nanoparticles, the core silica nanoparticles can be selectively removed with a base because the organosiloxane shell is highly stable, even in the high-pH solution [59]. Furthermore, it was reported that nanoparticles containing both silica and organosiloxane moieties could be transformed into a hollow structure. For example, the simple addition of organically bridged alkoxysilane to silica nanoparticles leads to the dissolution of silica and the subsequent formation of hollow-structured MSNs [60]. Recently, such hollow-structured nanoparticles with a high colloidal stability were prepared by covering silica nanoparticles with an organosiloxane shell, and the formation mechanism of the nanoparticles was investigated. The core MSNs were not dissolved in the solutions but were finally incorporated in the shell during the transformation process; the subsequent reconstruction of the shell part has been suggested (Fig. 4) [61].

Recently, anisotropic MSNs, including Janus-type nanoparticles, have also attracted great attention. Zhao et al. prepared Janus-type MSNs through an anisotropic growth of MSNs using bridged organoalkoxysilanes [31]. Such Janus nanoparticles contain two different kinds of guest species in two types of (meso)pores with different surface functional groups, and their

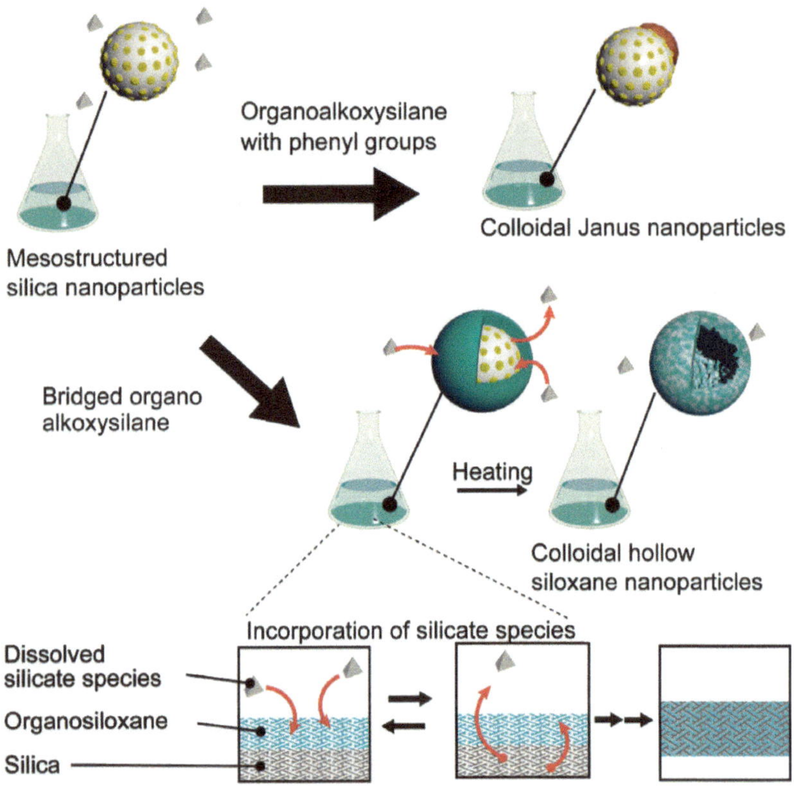

Fig. 4 Control of morphology using organoalkoxysilanes.

release can be controlled; this can be useful for drug delivery. On the other hand, we prepared Janus-type MSNs by simply adding phenyltriethoxysilane to a colloidal solution of MSNs. In this system, aromatic groups should be essential for the anisotropic growth of MSNs because core–shell structured nanoparticles were formed when we used other types of organoalkoxysilanes (Fig. 4) [27].

6. CONCLUSIONS

Various preparation methods for colloidal MSNs have been developed, and the size, morphology, and composition can be controlled appropriately. However, the integral control of these characteristics still remains a challenge because of their complex formation mechanisms. The development of the preparation methods for novel MSNs, and studies on the basic molecular-scale structure and formation mechanisms are important for the further progress of MSNs. An understanding of the structure, composition, and formation mechanisms would not only promote the progress into the material chemistry of mesoporous materials but would also enable the more effective use of these materials in biomedical applications.

REFERENCES

[1] T. Yanagisawa, T. Shimizu, K. Kuroda, C. Kato, Bull. Chem. Soc. Jpn. 63 (1990) 988–992.
[2] C.T. Kresge, M.E. Leonowicz, W.J. Roth, J.C. Vartuli, J.S. Beck, Nature 359 (1992) 710–712.
[3] M. Grün, I. Lauer, K.K. Unger, Adv. Mater. 9 (1997) 254–257.
[4] B.G. Trewyn, S. Giri, I.I. Slowing, V.S.Y. Lin, Chem. Commun. 31 (2007) 3236–3245.
[5] B.G. Trewyn, I.I. Slowing, S. Giri, H.T. Chen, V.S.Y. Lin, Acc. Chem. Res. 40 (2007) 846–853.
[6] I.I. Slowing, J.L. Vivero-Escoto, C.W. Wu, V.S.Y. Lin, Adv. Drug Deliv. Rev. 60 (2008) 1278–1288.
[7] K.K. Cotí, M.E. Belowich, M. Liong, M.W. Ambrogio, Y.A. Lau, H.A. Khatib, J.I. Zink, N.M. Khashab, J.F. Stoddart, Nanoscale 1 (2009) 16–39.
[8] J.M. Rosenholm, C. Sahlgren, M. Lindén, Nanoscale (2) (2010) 1870–1883.
[9] J.L. Vivero-Escoto, I.I. Slowing, V.S.Y. Lin, B.G. Trewyn, Small 6 (2010) 1952–1967.
[10] A. Popat, S.B. Hartono, F. Stahr, J. Liu, S.Z. Qiao, G.Q. Lu, Nanoscale 3 (2011) 2801–2818.
[11] Z. Li, J.C. Barnes, A. Bosoy, J.F. Stoddart, J.I. Zink, Chem. Soc. Rev. 41 (2012) 2590–2605.
[12] Y.S. Lin, K.R. Hurley, C.L. Haynes, J. Phys. Chem. Lett. 3 (2012) 364–374.
[13] Y. Chen, H. Chen, J. Shi, Adv. Mater. 25 (2013) 3144–3176.
[14] C.E. Fowler, D. Khushalani, B. Lebeau, S. Mann, Adv. Mater. 13 (2001) 649–652.
[15] Q. Cai, Z.-S. Luo, W.-Q. Pang, Y.-W. Fan, X.-H. Chen, F.-Z. Cui, Chem. Mater. 13 (2001) 258–263.
[16] E. Yamamoto, K. Kuroda, Bull. Chem. Soc. Jpn. 89 (2016) 501–539.
[17] S.H. Wu, C.Y. Mou, H.P. Lin, Chem. Soc. Rev. 42 (2013) 3862–3875.
[18] V. Valtchev, L. Tosheva, Chem. Rev. 113 (2013) 6734–6760.
[19] N. Suzuki, M.B. Zakaria, Y.D. Chiang, K.C.W. Wu, Y. Yamauchi, Phys. Chem. Chem. Phys. 14 (2012) 7427–7432.
[20] N. Suzuki, Y. Kamachi, K. Takai, S. Kiba, Y. Sakka, N. Miyamoto, Y. Yamauchi, Eur. J. Inorg. Chem. 17 (2014) 2773–2778.
[21] M.C. Chao, H.P. Lin, C.Y. Mou, B.W. Cheng, C.F. Cheng, Catal. Today 97 (2004) 81–87.
[22] S. Murai, S. Yao, T. Nakamura, T. Kawamoto, K. Fujita, K. Yano, K. Tanaka, Appl. Phys. Lett. 101 (2012) 151121.
[23] L. Bai, Z. Xie, K. Cao, Y. Zhao, H. Xu, C. Zhu, Z. Mu, Q. Zhong, Z. Gu, Nanoscale 6 (2014) 5680–5685.
[24] Y.S. Lin, C.P. Tsai, H.Y. Huang, C.T. Kuo, Y. Hung, D.M. Huang, Y.C. Chen, C.Y. Mou, Chem. Mater. 17 (2005) 4570–4573.
[25] T. Nakamura, H. Yamada, Y. Yamada, A. Gurtanyel, S. Hartmann, N. Husing, K. Yano, Langmuir 26 (2010) 2002–2007.
[26] K. Möller, J. Kobler, T. Bein, Adv. Funct. Mater. 17 (2007) 605–612.
[27] H. Ujiie, A. Shimojima, K. Kuroda, Chem. Commun. 51 (2015) 3211–3214.

[28] L. Zhang, F. Zhang, W.-F. Dong, J.-F. Song, Q.-S. Huo, H.-B. Sun, Chem. Commun. 47 (2011) 1225–1227.
[29] M. Xuan, J. Shao, X. Lin, L. Dai, Q. He, ChemPhysChem 15 (2014) 2255–2260.
[30] X. Ma, K. Hahn, S. Sanchez, J. Am. Chem. Soc. 137 (2015) 4976–4979.
[31] X. Li, L. Zhou, Y. Wei, A.M. El-Toni, F. Zhang, D. Zhao, J. Am. Chem. Soc. 136 (2014) 15086–15092.
[32] T. Suteewong, H. Sai, R. Hovden, D. Muller, M.S. Bradbury, S.M. Gruner, U. Wiesner, Science 340 (2013) 337–341.
[33] X. Fang, X. Zhao, W. Fang, C. Chen, N. Zheng, Nanoscale 5 (2013) 2205–2218.
[34] C. Urata, Y. Aoyama, A. Tonegawa, Y. Yamauchi, K. Kuroda, Chem. Commun. (2009) 5094–5096.
[35] S.P. Hudson, R.F. Padera, R. Langer, D.S. Kohane, Biomaterials 29 (2008) 4045–4055.
[36] Q. He, Z. Zhang, Y. Gao, J. Shi, Y. Li, Small 5 (2009) 2722–2729.
[37] V. Cauda, C. Argyo, D.G. Piercey, T. Bein, J. Am. Chem. Soc. 133 (2011) 6484–6486.
[38] Y.S. Lin, N. Abadeer, K.R. Hurley, C.L. Haynes, J. Am. Chem. Soc. 133 (2011) 20444–20457.
[39] H. Yamada, C. Urata, H. Ujiie, Y. Yamauchi, K. Kuroda, Nanoscale 5 (2013) 6145–6153.
[40] Y.S. Lin, C.Y.D. Lu, Y. Hung, C.Y. Mou, ChemPhysChem 10 (2009) 2628–2632.
[41] N.Z. Knezevic, J.O. Durand, Nanoscale 7 (2015) 2199–2209.
[42] X. Du, S.Z. Qiao, Small 11 (2015) 392–413.
[43] X. Du, B. Shi, J. Liang, J. Bi, S. Dai, S.Z. Qiao, Adv. Mater. 25 (2013) 5981–5985.
[44] V. Polshettiwar, D. Cha, X. Zhang, J.M. Basset, Angew. Chem. Int. Ed. 49 (2010) 9652–9656.
[45] A.B.D. Nandiyanto, S.G. Kim, F. Iskandar, K. Okuyama, Microporous Mesoporous Mater. 120 (2009) 447–453.
[46] K.-C. Kao, C.-Y. Mou, Microporous Mesoporous Mater. 169 (2013) 7–15.
[47] D. Shen, J. Yang, X. Li, L. Zhou, R. Zhang, W. Li, L. Chen, R. Wang, F. Zhang, D. Zhao, Nano Lett. 14 (2014) 923–932.
[48] H. Ishii, T. Ikuno, A. Shimojima, T. Okubo, J. Colloid Interface Sci. 448 (2015) 57–64.
[49] H. Yamada, H. Ujiie, C. Urata, E. Yamamoto, Y. Yamauchi, K. Kuroda, Nanoscale 7 (2015) 19557–19567.
[50] E. Yamamoto, S. Mori, A. Shimojima, H. Wada, K. Kuroda, Nanoscale 9 (2017) 2464–2470.
[51] J. Lu, M. Liong, Z. Li, J.I. Zink, F. Tamanoi, Small 6 (2010) 1794–1805.
[52] C.-Y. Lai, B.G. Trewyn, D.M. Jeftinija, K. Jeftinija, S. Xu, S. Jeftinija, V.S.Y. Lin, J. Am. Chem. Soc. 125 (2003) 4451–4459.
[53] J. Kobler, K. Möller, T. Bein, ACS Nano. 2 (2008) 791–799.
[54] V. Cauda, A. Schlossbauer, J. Kecht, A. Zürner, T. Bein, J. Am. Chem. Soc. 131 (2009) 11361–11370.
[55] C. Argyo, V. Weiss, C. Bräuchle, T. Bein, Chem. Mater. 26 (2014) 435–451.
[56] Y. Hoshikawa, H. Yabe, A. Nomura, T. Yamaki, A. Shimojima, T. Okubo, Chem. Mater. 22 (2010) 12–14.
[57] J. Kim, H.S. Kim, N. Lee, T. Kim, H. Kim, T. Yu, I.C. Song, W.K. Moon, T. Hyeon, Angew. Chem. Int. Ed. 47 (2008) 8438–8441.
[58] T. Ikuno, A. Nomura, K. Iyoki, A. Sugawara-Narutaki, T. Okubo, A. Shimojima, Chem. Lett. 42 (2013) 316–317.
[59] N. Koike, T. Ikuno, T. Okubo, A. Shimojima, Chem. Commun. 49 (2013) 4998–5000.
[60] Y. Yang, J. Liu, X. Li, X. Liu, Q. Yang, Chem. Mater. 23 (2011) 3676–3684.
[61] E. Yamamoto, S. Uchida, A. Shimojima, H. Wada, K. Kuroda, Chem. Mater. 30 (2018) 540–548.

CHAPTER TWO

Synthesis and Optical Applications of Periodic Mesoporous Organosilicas

Shinji Inagaki[1]
Toyota Central R&D Labs., Inc., Nagakute, Japan
[1]Corresponding author: e-mail address: inagaki@mosk.tytlabs.co.jp

Contents

Abstract

Periodic mesoporous organosilicas (PMOs), synthesized via surfactant-directed self-assembly of a polysilylated organic precursor ($R[Si(OR')_3]_n$; $n \geq 2$, R: organic group), are promising candidates such as catalysts and adsorbents, and for use in optical and electrical devices, owing to their high surface area, well-defined nanoporous structure, and highly functional organosilica framework. Their framework functionality can be widely tuned by selecting appropriate organic groups and controlling their arrangement. This chapter describes the synthesis and structure of PMOs with simple organic groups such as ethane and benzene, and the unique properties and optical applications of functional PMOs. Special light-harvesting properties and their exploitation in photocatalysis, highly emissive PMOs and their application to color-tunable transparent films, hole-transporting PMOs and their use in organic solar cells, and PMOs containing chelating ligands and their use as solid supports for heterogeneous metal complex catalysis are described.

The Enzymes, Volume 44
ISSN 1874-6047
https://doi.org/10.1016/bs.enz.2018.09.004
© 2018 Elsevier Inc.
All rights reserved.

1. INTRODUCTION

Ordered mesoporous silicas, prepared by surfactant-directed self-assembly, have great advantages, such as relatively large pores (2–30 nm), high specific surface area ($\sim$1000 $m^2\,g^{-1}$), and high thermal stability ($\sim$900 °C) [1,2]. However, the limited functionality of the silica framework narrows the range of applications of mesoporous silicas. To enhance functionality, much effort has been devoted to expanding the diversity of the framework composition of mesoporous materials. The framework composition has been extended from pure silica to a whole range of inorganic materials, including metallosilicates, non-Si transition metal oxides, metals, and carbons. In 1996, organic-silica hybrid mesoporous materials were synthesized by co-condensation of a monosilylated organic precursor [$R–Si(OR')_3$, R: organic group, R′: Me, Et, etc.] and a pure silica precursor [$Si(OR')_4$] (Fig. 1 (1)) [3]. The synthesis of mesoporous materials from a [$R–Si(OR')_3$] precursor alone is usually difficult. Rather, synthesis typically requires the addition of a large amount of the [$Si(OR')_3$] precursor (typically >75 mol% of total precursors) to form a stable framework. In hybrid mesoporous materials, organic groups

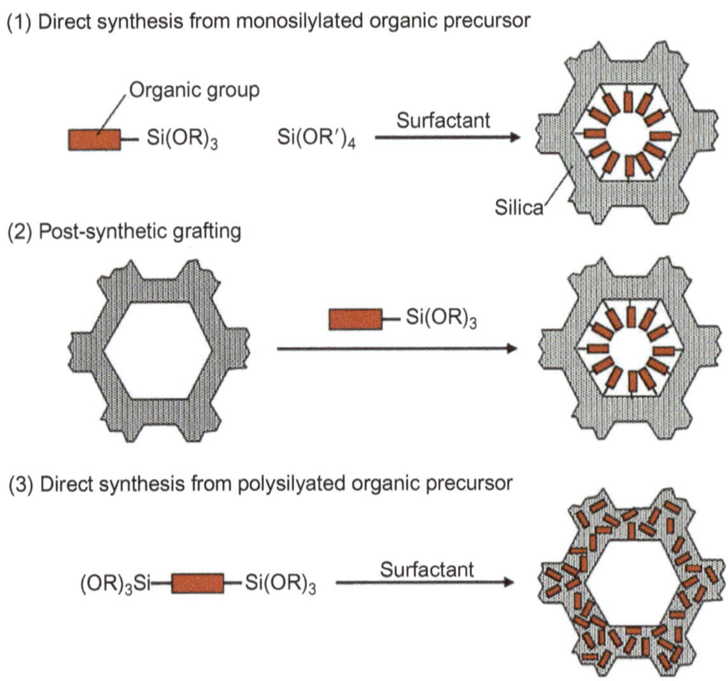

Fig. 1 Classification of organic-silica hybrid mesoporous materials.

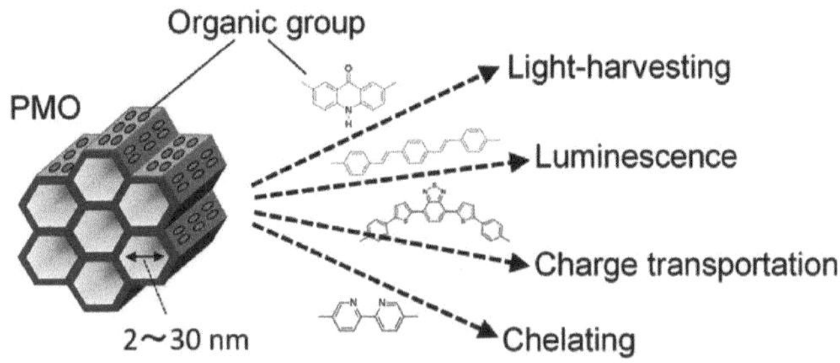

Fig. 2 Various functions which can be added on the PMOs framework.

are mainly grafted on pore surfaces, similar to organic-modified mesoporous silicas prepared by post-synthetic grafting of organic groups to mesoporous silica (Fig. 1 (2)). In 1999, the research groups of Inagaki [4], Stein [5], and Ozin [6] independently reported the synthesis of mesoporous organosilicas from bis-silylated organic precursors [$(R'O)_3Si$–R–$Si(OR')_3$, R = ethane and ethylene] (Fig. 1 (3)). Mesoporous materials derived from polysilylated organic precursors ($R[Si(OR')_3]_n$; $n \geq 2$) were termed periodic mesoporous organosilica (PMO). PMOs can be prepared from 100% polysilylated organic precursor or a mixture of polysilylated organic precursor and $Si(OR')_4$. Organic groups are distributed homogeneously inside the pore walls, which differs completely from organic-silica hybrid mesoporous materials prepared from a monosilylated organic precursor.

PMOs offer the great advantage of allowing the addition of various functions to the framework, such as light-harvesting, luminescence, charge transportation, and chelating properties, through the selection of the appropriate organic group (R) (Fig. 2). Highly functional PMOs can find new applications that are impossible to achieve with mesoporous silica. This chapter describes the synthesis of PMOs with simple organic groups such as ethane and benzene, and the unique properties and applications of highly functional PMOs. The focus is on optical applications such as light-harvesting photocatalysis, color-tunable luminescence films, photovoltaic devices, and heterogeneous metal complex catalysis.

2. SYNTHESIS OF PERIODIC MESOPOROUS ORGANOSILICAS

2.1 PMOs With Ethane Linker

PMOs containing ethane groups in their framework have been studied most extensively because ethane-bridged organosilane precursors are commercially available. Ethane-bridged PMOs are synthesized using various

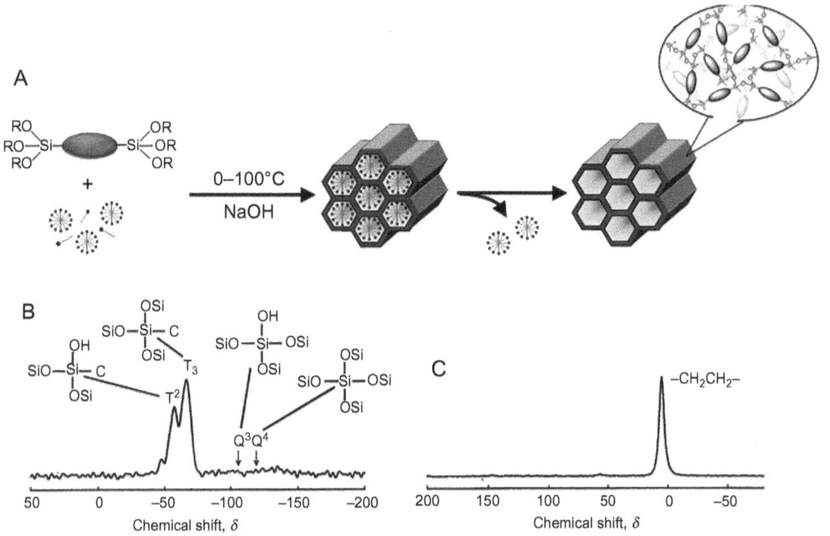

Fig. 3 (A) Surfactant-directed synthesis of ethane-PMO. (B) ^{29}Si MAS NMR and (C) ^{13}C CP MAS NMR spectra of ethane-PMO.

surfactants including alkyltrimethylammonium and block copolymers under basic and acidic conditions. Typically, ethane-bridged PMOs are prepared from 100% 1,2-bis(trimethoxysilyl)ethane in the presence of alkyltrimethylammonium under basic conditions (Fig. 3A) [4]. The surfactant template can be easily removed by extraction using an ethanol solution containing hydrochloric acid. The obtained ethane-PMOs show characteristic signals of T_n [$SiC(OSi)_n(OH)_{3-n}$, $n=1$–3] because, as revealed through ^{29}Si MAS NMR spectra (Fig. 3B), silicon species form single covalent bond with carbon. PMOs prepared from 100% bridged organosilane show only T_n signals. However, the occurrence of Q_n signals [$Si(OSi)_n(OH)_{4-n}$, $n=1$–4] suggests the cleavage of Si–C bonds during synthesis. Co-condensation of the bridged organosilica and pure silica precursors results in PMOs having a mixture of T_n and Q_n signals. The ^{13}C CP MAS NMR spectrum of ethane-PMO shows a single signal due to ethane carbons (Fig. 3C).

Three mesophases with a two-dimensional (2D) hexagonal (*P6mm*), three-dimensional (3D) hexagonal ($P6_3/mmc$), and cubic ($Pm-3n$) structure could be prepared using alkyltrimethylammonium as a surfactant under basic conditions by changing the synthesis temperature (t) and alkyl chain length of the surfactant (n) (Fig. 4) [4,7]. The 2D hexagonal and cubic $Pm-3n$ mesophases were obtained at $n=18$ and 16, respectively, at the same temperature (95 °C). Interestingly, a small difference in alkyl chain length of the

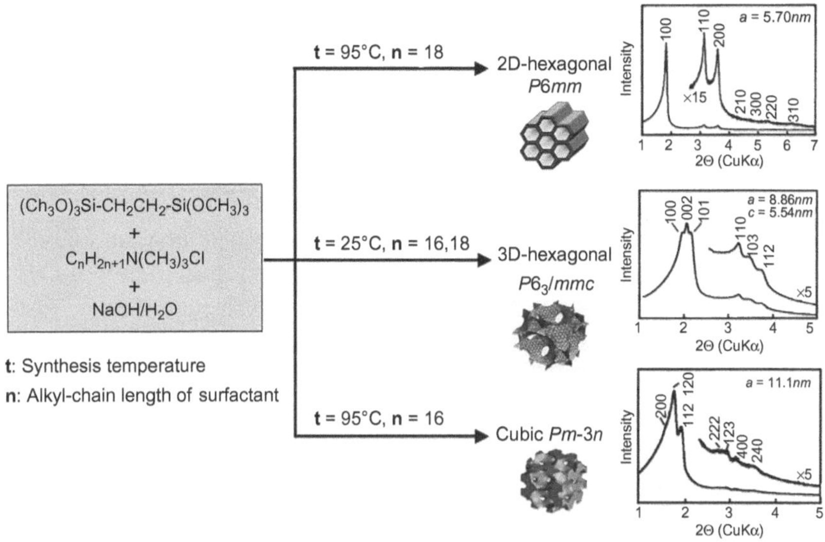

Fig. 4 Synthesis conditions for three mesophases of ethane-PMOs and their XRD patterns.

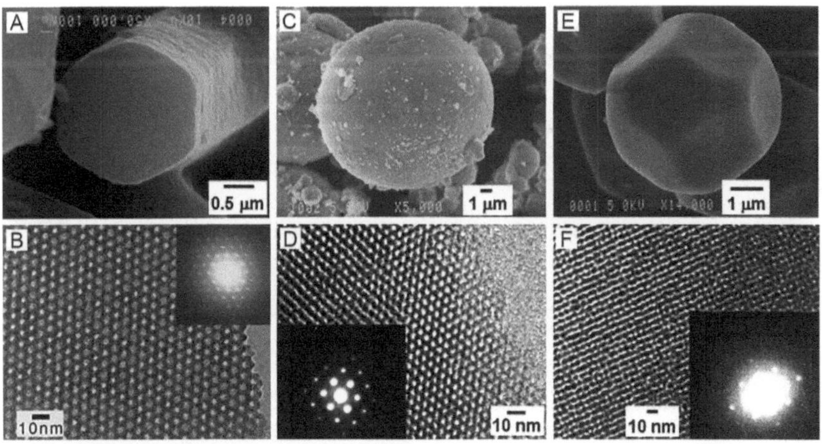

Fig. 5 Scanning (A, C, E) and transmission (B, D, F) electron micrographs of ethane-PMOs with 2d-hexagonal *P6mm* (A, B), 3d-hexagonal *P6$_3$/mmc* (C, D) and cubic *Pm* − 3*n* (E, F).

surfactant leads to different mesophases. The XRD patterns and diffraction peak assignments are shown in Fig. 4.

Scanning and transmission electron micrographs of ethylene-PMOs are shown in Fig. 5. Ethane-PMOs show well-defined external morphologies with hexagonal rod (2D-hexagonal), spherical (3D-hexagonal) and decaoctahedral (cubic) shapes. These shapes reflect the symmetries of the underlying arrangement of mesopores, as shown in the TEM images.

2.2 Crystal-Like PMOs With Benzene Linker

The pore walls of ethane-PMOs are amorphous, although the mesopores show a regular arrangement. However, a benzene-PMO prepared from a 100% benzene-bridged organosilane precursor under basic conditions exhibits a crystal-like pore wall structure in addition to an ordered pore arrangement (Fig. 6A) [8]. The XRD pattern for the benzene-PMO showed some peaks at a medium scattering angle assignable to a lamellar structure with a basal spacing of 7.6 Å in addition to low-angle reflections due to the ordered mesostructure (Fig. 6B). A lamellar periodicity of 7.6 Å was determined by high-resolution TEM (Fig. 6C). The TEM image reveals many lattice fringes with a spacing of 7.6 Å on the pore walls in the direction perpendicular to the channels. Electron diffraction spots indicating a periodicity of 7.6 Å are observed in the same direction. These observations clearly indicate the presence of molecular-scale periodicity in the pore walls of benzene-PMO. Fig. 7 shows the structural model of a benzene-PMO. The benzene-PMO has a 2D-hexagonal mesoporous structure with a pore diameter of 3.8 nm and alternating phenylene and silica layers in its pore walls with a periodicity of 7.6 Å. The nitrogen adsorption–desorption isotherm for the benzene-PMO is a typical type IV isotherm, indicating the presence of uniform mesopores, as confirmed by the narrow pore size distribution curve.

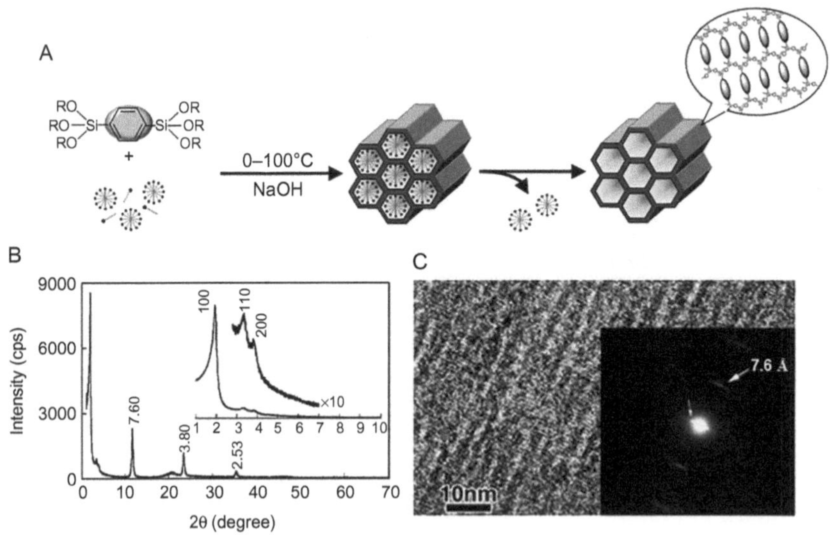

Fig. 6 (A) Surfactant-directed synthesis of a crystal-like benzene-PMO. (B) XRD pattern and (C) TEM image of benzene-PMO.

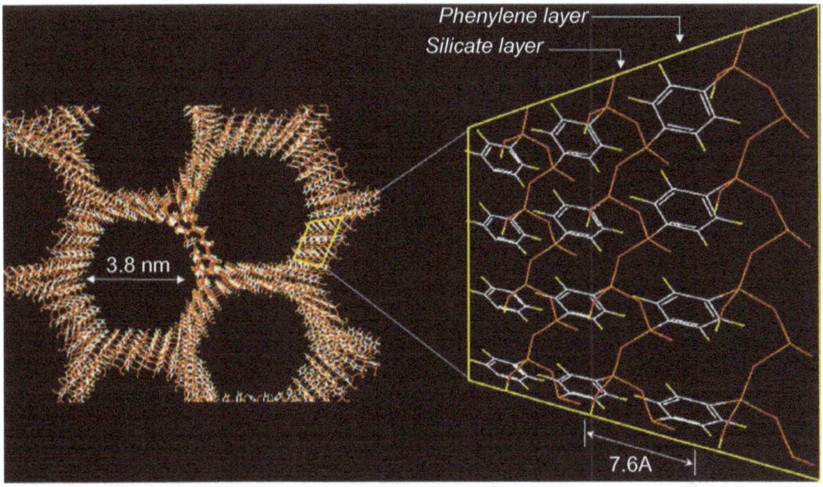

Fig. 7 Structural model of a crystal-like benzene-PMO.

The lamellar arrangement of benzene and silica moieties was formed by the hydrophobic–hydrophilic interaction of $(HO)_3Si–Ph–Si(OH)_3$ species in an aqueous solution. $(EtO)_3Si–Ph–Si(OEt)_3$ is hydrolyzed to $(HO)_3Si–Ph–Si(OH)_3$ under basic conditions. $(HO)_3Si–Ph–Si(OH)_3$ self-assembles into a lamellar aggregate in the aqueous solution owing to the strong hydrophobic–hydrophilic interaction. The benzene rings in the PMO framework can rotate around the Si–C axis, as confirmed by variable-temperature solid-state ^2H spin-echo NMR measurements [9]. This result indicates that the driving force for self-assembly is not π–π interactions of the benzene rings, but the hydrophobic–hydrophilic interaction of $(HO)_3Si–Ph–Si(OH)_3$. No molecular-scale periodicity is observed in the pore walls of benzene-PMOs prepared under acidic conditions [10].

2.3 Other Crystal-Like PMOs

A similar molecular-scale periodicity was observed for PMOs containing other organic groups such as biphenyl [11], 1,3-benzene [12], ethylene [13], naphthalene [14], divinylbenzene [15], divinylpyridine [16], phenylpyridine [17], and bipyridine [18] (Table 1). These crystal-like PMOs have a lamellar structure with a periodicity of 5.6–11.9 Å, depending on the length of the bridging organic groups. Rigid, hydrophobic bridging organic groups are appropriate for the formation of a lamellar arrangement. However, it was difficult to obtain a crystal-like PMO from organosilane precursors containing longer bridging organics such as anthracene.

Table 1 Molecular-Scale Periodicity of Crystal-Like PMOs With a Lamellar Structure

Organic Group	Molecular-Scale Periodicity (Å)
	7.6
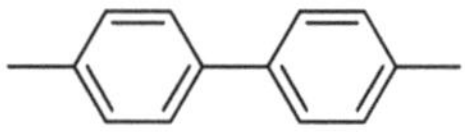	11.7
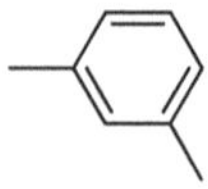	7.6
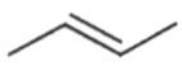	5.6
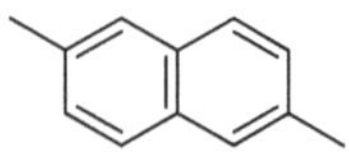	9.4
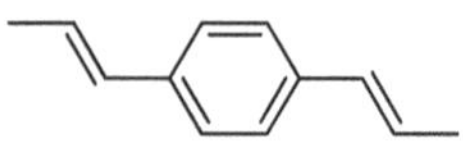	11.9
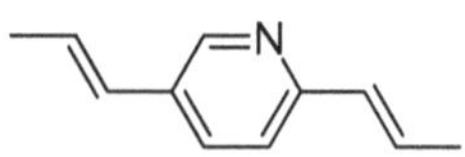	11.6
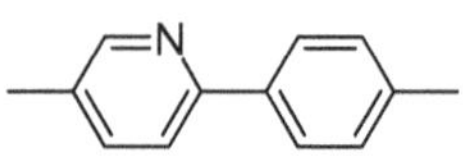	11.7
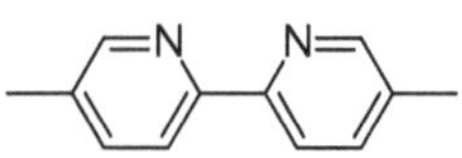	11.6

Different strategies were employed to construct crystal-like PMOs containing large organic groups. PMOs containing perylene bisimide in their frameworks showed a molecular-scale periodicity of around $d = 0.35$ nm, corresponding to the typical π–π stacking distance [19]. The key to the formation of this type of new ordered structure is full utilization of the π–π stacking nature of the hydrophobic PBI moieties, rather than the hydrophobic–hydrophilic interaction. An electron spin resonance (ESR) study of a perylene bisimide-containing PMO-doped with hydrazine suggested hopping or delocalization of unpaired electrons over several electron-accepting PBI sites. The formation of molecular-scale periodicity in the pore walls of *PMOs* also involved hydrogen bonding. PMOs containing cyclohexane triamide moieties exhibited molecular-scale periodicity assignable to the 2D hexagonal symmetry in addition to the ordered mesoporous structure [20]. The pore walls are composed of H-bonded columns of cyclohexane triamides aligned parallel to the channels. H-bonded PMOs have the advantage of being able to accommodate functional guest molecules in their frameworks through hydrogen-bonding interactions.

3. FUNCTIONS AND APPLICATIONS

3.1 Light-Harvesting Properties

A light-harvesting antenna in natural photosynthesis has the important role of promoting multielectron reactions through the concentration of sunlight with low photon density. The excited state of the chlorophyll molecules in the antenna, which is induced by light absorption, is transferred to neighboring chlorophyll molecules by resonance energy transfer, and finally funneled to the reaction center (special pair). The excitation frequency for the reaction center can be enhanced 100- to 200-fold because the antennas in plants usually have 100–200 chlorophyll molecules per reaction center. PMOs exhibit excellent light-harvesting ability, comparable to that observed in photosynthesis. The organic groups in the PMO framework absorb light and transfer the excitation energy to the dopant dye molecules in the mesopores (Fig. 8A) [21]. Significant enhancement in the emission from the coumarin dye in biphenyl-PMO was observed when it was excited by 280-nm light, as compared to 380-nm light (Fig. 8B). The 280-nm light was absorbed by the many biphenyl groups in the PMO framework and the excitation energy was transferred efficiently to the coumarin dye molecules. It was confirmed that the quantum efficiency of the energy transfer was almost 100%. This high efficiency was attributable to the unique structure of the PMO. The biphenyl groups in the PMO framework surround a coumarin molecule, and the intermolecular distances between biphenyl and the coumarin dye are within the Förster radius (3.2 nm). An experiment in which the amount of coumarin dopant was varied suggested that the excitation energy incident on the 125 biphenyl groups was funneled to a single coumarin molecule. Light-harvesting properties are observed for visible-light-absorptive PMOs such as acridone-PMO [22].

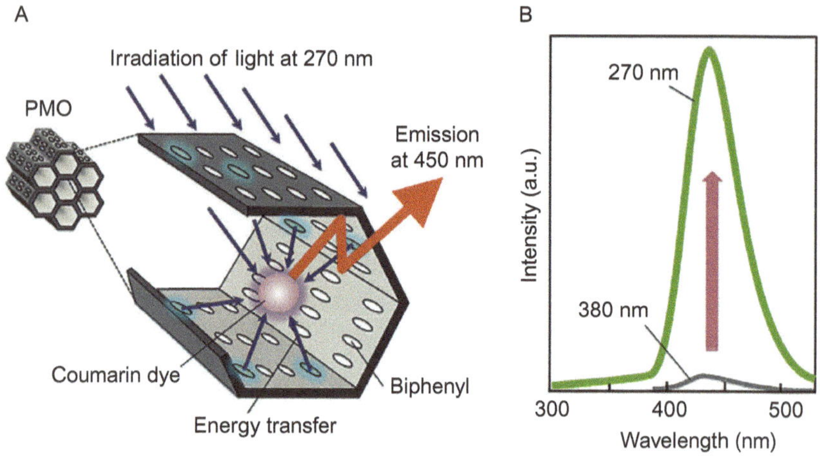

Fig. 8 (A) Mechanism of light-harvesting by PMO and (B) fluorescence spectra of coumarin dye-doped biphenyl-PMO excited at 270 and 380 nm.

3.2 Light-Harvesting PMO Photocatalysis

Photocatalytic performance can be enhanced by immobilization of the photocatalyst in the mesopores of a light-harvesting PMO. One example of such light-harvesting PMO photocatalysis involves the rhenium tricarbonyl complex, $[Re(bpy)_3(CO)_3Cl]$, immobilized on Bp-PMO [23]. The fluorescence emission band for Bp-PMO overlaps well with the absorption band for the Re complex. Immobilization of the Re complex in the mesopores quenched the fluorescence from Bp-PMO, suggesting efficient energy transfer from biphenyl groups in the PMO framework to Re complexes in the mesopores. The Re complex immobilized on Bp-PMO was dispersed in acetonitrile solution containing a sacrificial agent (triethanolamine: TEOA) and irradiated with 280- and 365-nm light. When irradiated at 365 nm, a small amount of CO was evolved, while a larger amount of CO (4.4 times larger) was generated under irradiation at 280 nm. The 280-nm light was efficiently absorbed by the highly concentrated biphenyl groups in Bp-PMO, and the excitation energy was transferred to the Re complex. The 365-nm light could not be absorbed by biphenyl groups; rather, it was absorbed by the low-concentration Re complexes. The excitation frequency for the Re complex is much enhanced by the light-harvesting properties of the PMO.

A visible-light-harvesting PMO photocatalyst was also developed. Acridone-PMO (Acd-PMO) can absorb visible light up to 450 nm and emit

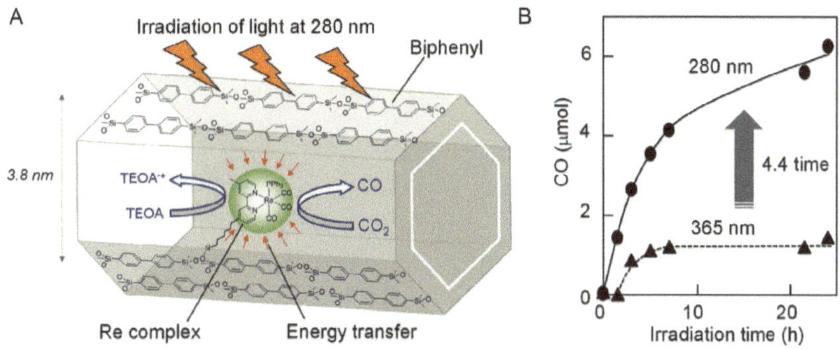

Fig. 9 (A) Light-harvesting photocatalysis for CO_2 reduction for Re complex immobilized on Bp-PMO. (B) CO generation from CH_3CN solution containing Re/Bp-PMO and TEOA under a CO_2 atmosphere by irradiation of light at 280 and 365 nm.

at 520 nm [22]. A Ru–Re complex is an efficient homogeneous photocatalyst for CO_2 reduction, with an absorption band at 450–550 nm, which overlaps the emission band of Acd-PMO. The Ru–Re dinuclear complex, bearing methylphosphonic acids as anchor groups, was immobilized on Acd-PMO (Fig. 9) [24]. The Acd-PMO immobilized with the Ru–Re complex showed efficient photocatalysis for CO_2 reduction to form CO under irradiation at 405 nm. The acridone groups in the PMO framework absorbed the 405-nm light and transferred the excitation energy to the Ru–Re complexes in the mesochannels. The photocatalytic activity was 10 times higher than that of Ru–Re-complex-immobilized mesoporous silica with no light-harvesting function, revealing the light-harvesting effect of Acd-PMO.

A PMO containing $[Ru(bpy)_3]^{2+}$ in its framework (Ru-PMO) absorbed a wide range of visible light up to 600 nm [25]. Ru-PMO could be used as a solid photosensitizer for heterogeneous photocatalysis. A Ru-PMO loaded with IrO_x nanoparticles showed efficient photocatalysis for water oxidation in the presence of NaS_2O_8 as a sacrificial agent under irradiation at 450 nm (Fig. 10). The excited $[Ru(bpy)_3]^{2+}$ in the framework was oxidized by $S_2O_8{}^{2-}$ and extracted electrons from IrO_x nanoparticles. The electron-deficient IrO_x oxidized water to generate O_2. A very high quantum yield of 17.3% was obtained in this photocatalysis system. Ru-PMO can also be used as a solid photosensitizer for H_2 evolution from water in the presence of 1-benzyl-1,4-dihydronicotinamide (BNAH) after loading Pt nanoparticles on the pore surface [25].

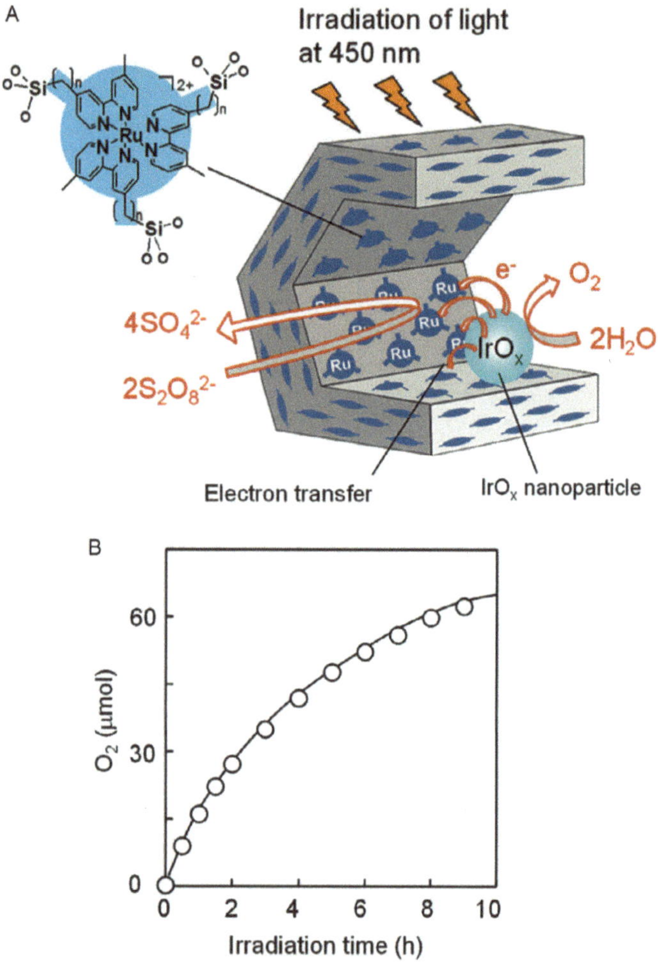

Fig. 10 (A) Photocatalytic water oxidation for IrO_x nanoparticle supported Ru complex-containing PMO. (B) O_2 generation from aqueous solution containing IrO_x/Ru-PMO and NaS_2O_8 under irradiation of light at 450 nm.

3.3 Highly Emissive and Color-Tunable Mesostructured Organosilica Films

Mesostructured organosilica thin films can be prepared by spin- or dip-coating a sol containing ethanol, an organosilane precursor, a surfactant, and a small amount of water and hydrochloric acid (Fig. 11). Highly emissive mesostructured films were prepared from organosilane precursors containing π-conjugated organic groups, such as 2-ethylhexyloxy-substituted

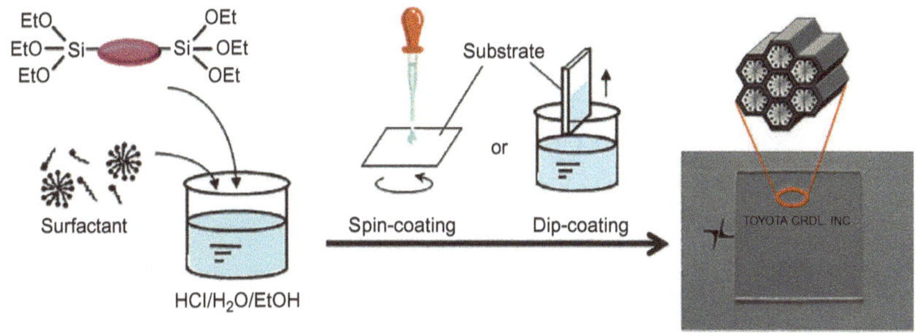

Fig. 11 Preparation of a mesostructured organosilica thin film using spin- or dip-coating of sol solution.

oligo(phenylenevinylene) (OPV) [26], tetraphenylpyrene (TPPy) [27], and spirobifluorene (SBF) [28]. These transparent mesostructured films showed blue light emission with a high quantum efficiency of 59–79%. The π-conjugated organic groups were separated from each other due to the bulky substitutes (OPV or TPPy) or unique spatial conformation (SBF) in the framework, which reduced concentration quenching of the chromophores even at high density.

Mesostructured organosilica films have the advantage that the emission color can be tuned by doping the mesopore region, filled with surfactant micelles, with a dye. Fig. 12 illustrates dual emission from a mesostructured OPV film doped with a rubrene dye [29]. An undoped mesostructured OPV film exhibits strong blue light emission at 485 nm with a quantum yield of 61%. The rubrene dye was selected because its absorption band well overlaps the emission band for the mesostructured OPV film. The rubrene dye dopant quenched the blue light emission from the OPV units and generated yellow light emission from the rubrene dye because the excitation energy transfer occurred from the OPV units to the rubrene dye. The fraction of the energy transferred can be controlled by varying the doping amount of the rubrene dye. The blue light emission from mesostructured OPV film was almost completely quenched by 5–10 mol% rubrene doping, and strong yellow light emission with a quantum yield of 76% was observed. At a doping amount of 0.6 mol% of rubrene, both blue and yellow light were emitted, which resulted in pseudo white light emission with a quantum yield of 67%. A rhodamine 6G-doped mesostructured TPPy film also showed pseudo white light emission with a quantum yield of 43% [27]. The dye-doped TPPy film was subsequently used to develop a white-light-emitting diode.

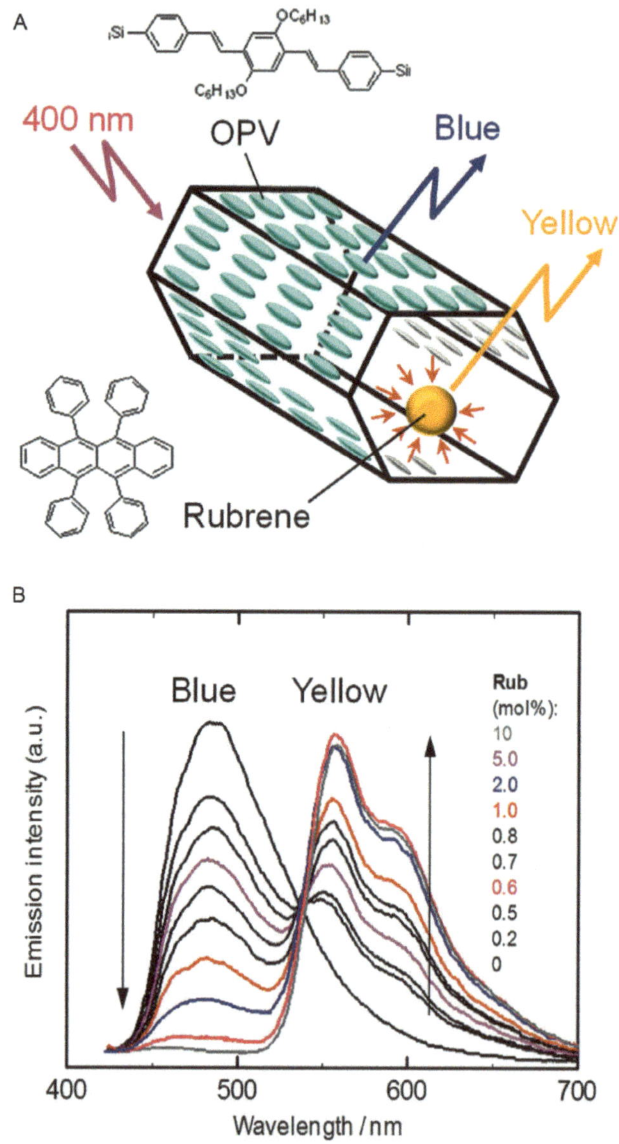

Fig. 12 (A) Dual emission from rubrene dye doped mesostructured OPV film. (B) Fluorescence spectra of rubrene-doped mesostructured OPV films with different doping amounts of rubrene excited at 400 nm.

3.4 Hole-Transporting PMO Films and Their Application to Organic Solar Cells

Hole-transporting properties can be imparted to the PMO pore walls by the introduction of hole-conductive organic groups into the framework. A PMO film was prepared from a three-armed phenylenevinylene compound with three triethoxysilyl groups by dip-coating (Fig. 13) [30]. The hole mobility for this PMO film, as measured by the time-of-flight (TOF) method, was 3.2×10^{-5} $cm^2\ V^{-1}\ s^{-1}$. This is comparable to those for amorphous films of poly(*p*-phenylenevinylene) polymers and lower than that for the typical hole-conductive polymer polythiophene (P3HT) (10^{-3} $cm^2\ V^{-1}\ s^{-1}$). The PMO film absorbed only UV light below a wavelength of 350 nm.

A visible-light-absorptive hole-transporting PMO film was also synthesized from an organosilane precursor containing 4,7-dithienyl-2,1,3-benzothiadiazole (DTBT) (Fig. 14A) [31]. The hole mobility for this DTBT-PMO film was 6×10^{-5} $cm^2\ V^{-1}\ s^{-1}$, as determined by the space-charge-limited current (SCLC) method. A solar cell device was fabricated using the DTBT-PMO film as a p-type active layer and [6,6]-phenyl C61 butyric acid methyl ester (PCBM) as a n-type layer (Fig. 14B). A buffer layer composed of a nonporous DTBT film was formed on a PEDOT:PSS/ITO anode to avoid direct contact between PCBM and the anode. A sol with the DTBT precursor and surfactant was then spin-coated on the buffer layer and the surfactant was removed by solvent extraction to form cage-like pores. Subsequently, the PCBM solution was cast on top of the film in order to infiltrate the mesopores. LiF and Al were deposited

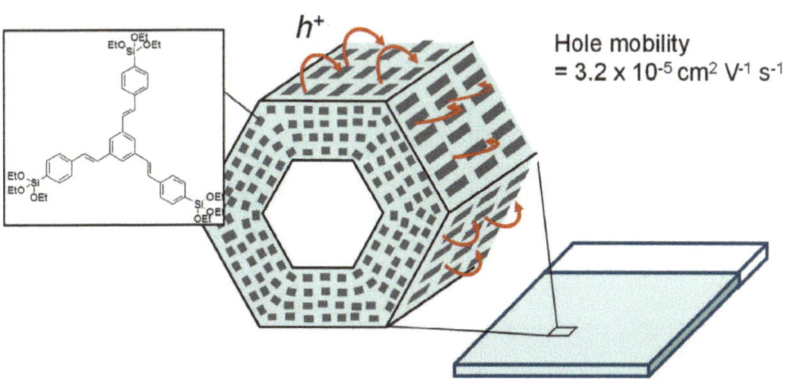

Fig. 13 Hole-transporting mesostructured thin film containing three-armed oligo (phenylenevinylene) groups in the framework.

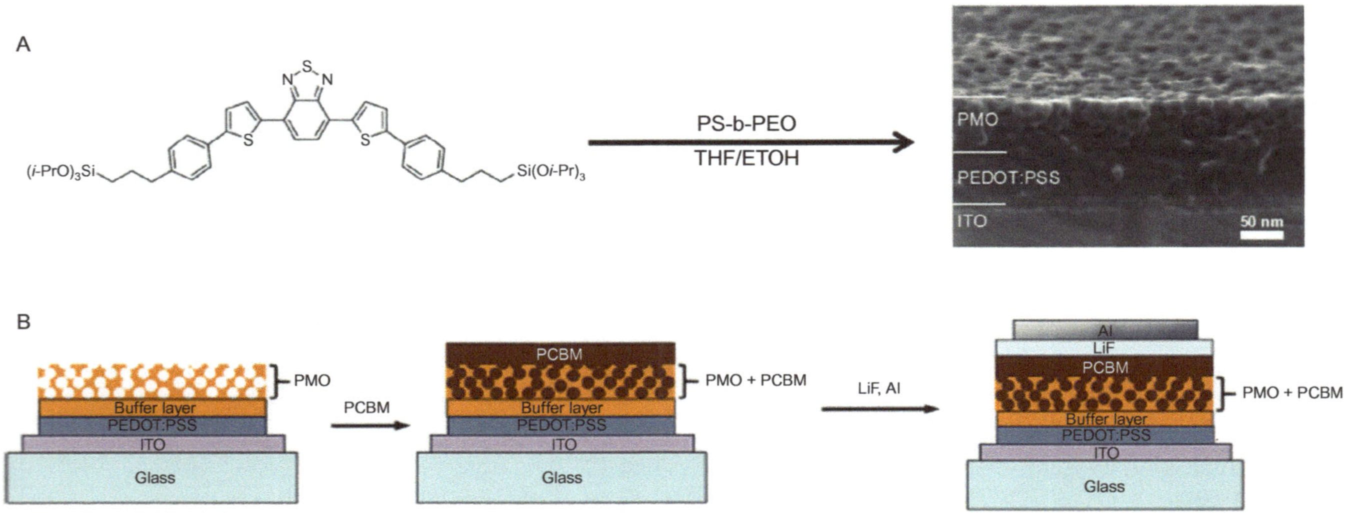

Fig. 14 (A) Preparation of DTBT-PMO film using spin-coating of sol solution followed by solvent extraction of surfactant. (B) Fabrication of a solar cell device by infiltration of PCBM solution into the mesopores of DTBT-PMO film.

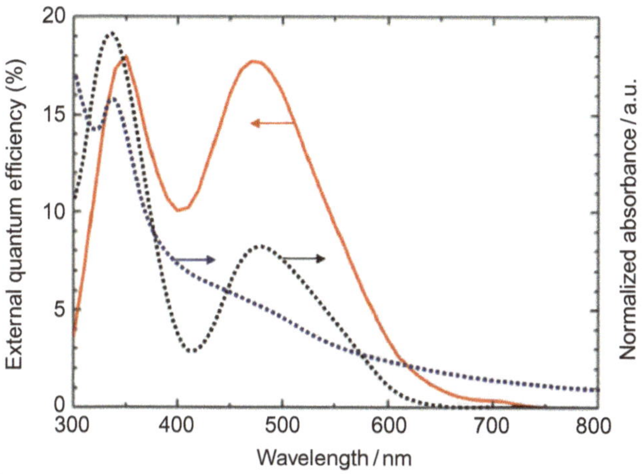

Fig. 15 External quantum efficiency of DTBT-PMO/PCBM device (red) and normalized absorbance of DTBT-PMO film (black) and PCBM solution (blue).

onto the PCBM layer as an electron extraction layer and cathode, respectively. Fig. 15 shows the external quantum efficiency of the DTBT-PMO/PCBM device and the normalized absorbance of the DTBT-PMO film and PCBM solution. The external quantum efficiency spectrum has two peaks at around 340 and 480 nm, which correspond to absorption by DTBT-PMO and PCBM, respectively, as shown in Fig. 15. This indicates that DTBT-PMO and PCBM act as p-type and n-type active layers in the solar cell device. The external quantum efficiency reached 18%, which is >500 times that for a conventional porphyrin-PMO/PCBM device [32]. The *J*–*V* characteristics of the DTBT-PMO/PCBM solar cell device revealed its power conversion efficiency (PCE = 0.61%), open circuit voltage (V_{oc} = 0.91 V), short circuit current density (JSC = 1.89 mA cm^{-2}), and fill factor (FF = 0.36). It is considered that the molecular design of the organosilane precursor and the packing control of the electro-active organic units in the PMO framework should increase the hole mobility of the PMO active layer to achieve further improvement in device performance.

3.5 Ligand-Containing PMOs and Heterogeneous Molecular Photocatalysis

Immobilization of metal complex catalysts is of great practical importance for easy recovery and recycling of metal catalysts, and is suitable for efficient flow-through reactions. The typical conventional approach for immobilization is grafting metal complexes on a solid surface using a molecular linker.

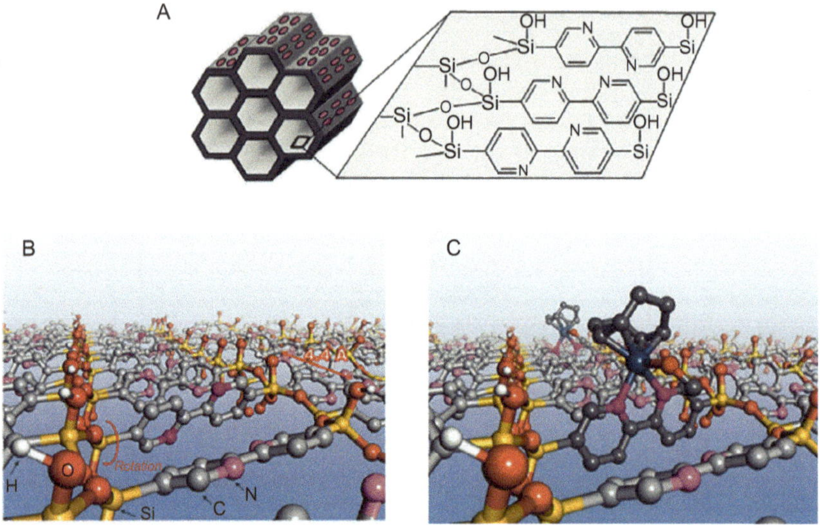

Fig. 16 (A) Schematic illustration of BPy-PMO. (B, C) CG-images of pore surface of BPy-PMO before (B) and after (C) immobilization of metal complexes.

However, this approach has some drawbacks, such as a decrease in activity and selectivity due to the unfavorable interaction between the metal center and solid surface. Metals are sometimes leached out, owing to the cleavage of molecular linkers. PMOs containing bipyridine ligands in its framework (BPy-PMO) represent a new type of solid support because metal complexes can be directly immobilized on the pore surfaces without using a molecular linker [18]. BPy-PMO has a unique pore wall structure in which bipyridine ligands are densely and regularly packed and exposed on the surface (Fig. 16A). The average bipyridine-bipyridine distance is 4.4 Å, which is longer than the typical π–π stacking distance of 3.5 Å (Fig. 16B). Thus, the framework bipyridine ligands can rotate around the Si–C axis and preserve the reactivity to form metal complexes. The framework bipyridine ligands can be used as one of the ligands of metal complexes, resulting in the formation of metal complexes on the pore surface of BPy-PMO (Fig. 16C).

The formation of a variety of metal complexes on BPy-PMO and their heterogeneous catalysis have been reported, for example, Ir(cod)(OMe)(BPy) (cod = cyclooctadiene, BPy = framework bipyridine) for direct C–H borylation of arenes [18,33], $Ru(CO)_2Cl_2$(BPy) for selective oxidation of alkanes [34] and photocatalytic CO_2 reduction to formate [35],

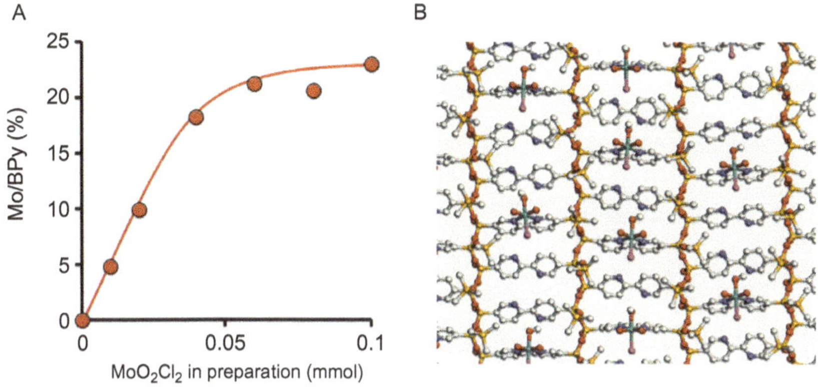

Fig. 17 (A) Relation between MoO_2Cl_2 in preparation and loaded amount of Mo (Mo/BPy). (B) CG-images of $MoO_2Cl(OH)$ complexes on the pore surface of BPy-PMO (Mo/BPysurface = 35%).

IrCp*Cl(BPy) (Cp* = pentamethylcyclopentadienyl) for chemical water oxidation [36], RhCp*Cl(BPy) for transfer hydrogenation of nitrogen heterocycles [37], $Re(CO)_3Cl(BPy)$ for photocatalytic CO_2 reduction to CO [38], and $MoO_2Cl_2(BPy)$ for epoxidation of olefins [39]. Metal complex formation on the pore surface of BPy-PMO was achieved by treatment of BPy-PMO powder with a solution containing a metal complex precursor. The loading amount of metal complexes can be controlled by changing the amount of metal complex precursor. Fig. 17A shows the relationship between the loading amount of Mo complex (Mo/BPy molar ratio) and the amount of MoO_2Cl_2 precursor [39]. It can be seen that the Mo/BPy ratio increased with the amount of MoO_2Cl_2, and saturated at Mo/BPy = 23%. Two-thirds of the bipyridine ligands are exposed on the pore surface because the pore walls of BPy-PMO are composed of three layers of BPy groups. Thus, Mo/$BPy_{surface}$ is 35%. The value indicates that Mo complexes are immobilized on the pore surface at intervals of two or three bipyridine ligands, as shown in the CG image (Fig. 17B).

$Re(CO)_3Cl(bpy)$ immobilized on BPy-PMO (Re/BPy-PMO) showed heterogeneous photocatalysis for CO_2 reduction to CO in the presence of 1,3-dimethyl-2-phenyl-2,3-dihydro-1*H*benzo[*d*]-imidazole (BIH) as a sacrificial agent under irradiation of 420-nm light [38]. Interestingly, the photocatalytic activity was enhanced by co-immobilization of $[Ru(bpy)_3]^{2+}$ on the pore surface of Re/BPy-PMO (Fig. 18). The efficient photocatalysis with the Ru/Re/BPy-PMO system suggests that electron transfer occurs from

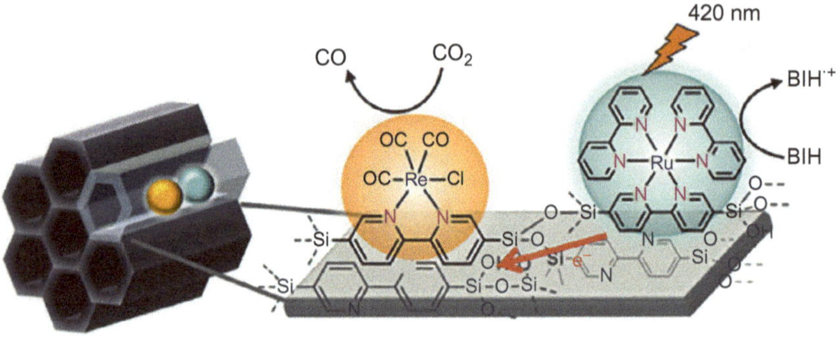

Fig. 18 Photocatalytic CO_2 reduction for Re and Ru complexes immobilized on BPy-PMO.

the Ru photosensitizer to the Re catalyst on the pore surface. Density functional theory (DFT) calculations were carried out to determine the molecular orbitals of Ru and Re complexes on PMO. The highest occupied molecular orbital (HOMO) was distributed around the Ru atom, whereas the second-lowest and lowest unoccupied molecular orbitals (LUMO) + 1 and LUMO were localized on the framework BPy ligands of the Ru and Re complexes, respectively. This indicates that an electron on a Ru atom is initially excited by irradiation, after which it transfers to the BPy framework, followed by tunneling from the Ru complex to the Re complex via through-Si–O–Si-bond interactions between the molecular orbitals in the pore walls. These findings should pave the way for the design of efficient heterogeneous CO_2 reduction photocatalysis systems.

Ir(cod)(OMe)(bpy) immobilized on BPy-PMO [Ir(cod)/BPy-PMO] exhibited higher catalytic performance for direct C–H borylation of benzene than conventional heterogeneous Ir catalysts grafted on silica gel (Ir-Silica gel) and mesoporous silica (Ir-MCM-41) using a molecular linker (Fig. 19) [18]. The Ir-grafted catalysts were deactivated during the reaction because of the unfavorable interaction between the metal center and the solid surface due to the flexible molecular linker. Furthermore, the homogeneous Ir catalyst [Ir(cod)(OMe)(bpy)] showed lower catalytic activity than Ir(cod)/BPy-PMO. The homogeneous Ir catalyst was also deactivated owing to the aggregation of Ir species during the reaction. However, Ir(cod)/BPy-PMO was not deactivated because the Ir complexes were fixed directly on the pore surfaces, which prevents aggregation and unfavorable interaction with the solid surface. Ir(cod)/BPy-PMO could be

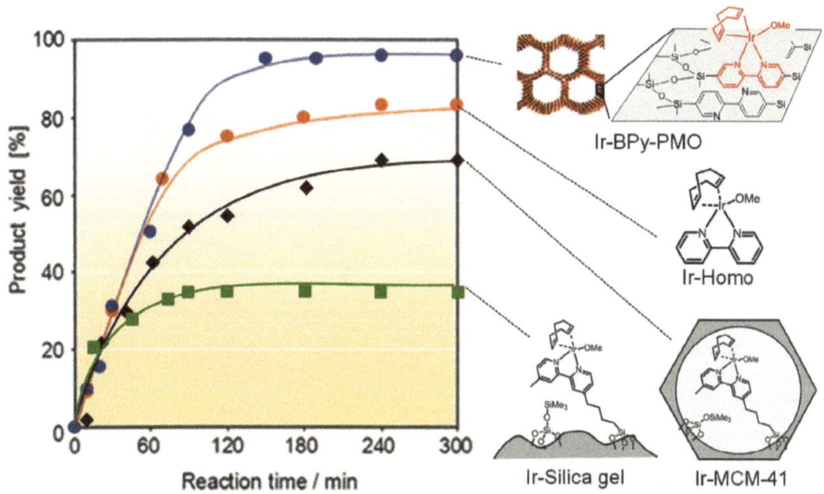

Fig. 19 Direct C–H borylation of benzene at 80 °C over Ir(cod)(OMe)(bpy) immobilized on BPy-PMO, silica gel and mesoporous silica (MCM-41), and homogeneous Ir complex (Ir-Homo).

easily recovered by filtration after the reaction. The recovered PMO catalyst retained good catalytic activity for at least three cycles, with only a slight loss of yield. In contrast, homogeneous and grafted catalysts showed almost no catalytic activity when recycled.

The initial reaction rate for Ir(cod)/BPy-PMO is almost the same as that for the homogeneous catalyst with the same amount of Ir (Fig. 19). The reaction rate should decrease for heterogeneous catalysts with small pores owing to the diffusion limitation of the reactant and product molecules. The comparable reaction rate suggests a very small diffusion limit on molecules in the mesochannels, because of the relatively large pore size for Ir(cod)/BPy-PMO (3.5 nm). A similarly high reaction rate was observed for the chemical water oxidation reaction using IrCp*-complex-immobilized BPy-PMO (IrCp*/BPy-PMO) [36]. This IrCp*/BPy-PMO catalyst showed high catalytic activity for water oxidation with a turnover frequency (TOF) of 2.8 min^{-1}. This value is comparable to that for the homogeneous Ir catalyst [IrCp*Cl(bpy)] (3.0 min^{-1}). An IrCp*-complex-immobilized metal-organic framework (MOF) showed a TOF that was one order of magnitude lower (0.12 min^{-1}) than that of IrCp*/BPy-PMO because the pore diameter of MOF was <1 nm [37]. This result points to the great advantage of PMO with its relatively large pores, which allow smooth mass transfer in the mesochannels.

REFERENCES

[1] T. Yanagisawa, T. Shimizu, K. Kuroda, C. Kato, The preparation of alkyltrimethylammonium-kanemite complexes and their conversion to microporous materials, Bull. Chem. Soc. Jpn. 63 (1990) 988–992. S. Inagaki, Y. Fukushima, K. Kuroda, Synthesis of highly ordered mesoporous materials from a layered polysilicate, J. Chem. Soc. Chem. Commun. (1993) 680–682.

[2] C.T. Kresge, M.E. Leonowicz, W.J. Roth, J.C. Vartuli, J.S. Beck, Ordered mesoporous molecular sieves synthesized by a liquid-crystal template mechanism, Nature 359 (1992) 710–712. J.S. Beck, J.C. Vartuli, W.J. Roth, M.E. Leonowicz, C.T. Kresge, K.D. Schmitt, C.T.-W. Chu, D.H. Olson, E.W. Sheppard, S.B. McCullen, J.B. Higgins, J.L. Schlenker, A new family of mesoporous molecular sieves prepared with liquid crystal templates, J. Am. Chem. Soc. 114 (1992) 10834–10843.

[3] D.J. Macquarrie, Direct preparation of organically modified MCM-type materials. Preparation and characterization of aminopropyl-MCM and 2-cyanoethyl-MCM, Chem. Commun. (1996) 1961–1962.

[4] S. Inagaki, S. Guan, Y. Fukushima, T. Ohsuna, O. Terasaki, Novel mesoporous materials with a uniform distribution of organic groups and inorganic oxide in their frameworks, J. Am. Chem. Soc. 121 (1999) 9611–9614.

[5] B.J. Melda, B.T. Holland, C.F. Blanford, A. Stein, Mesoporous sieves with unified hybrid inorganic/organic framework, Chem. Mater. 11 (1999) 3302–3308.

[6] T. Asefa, M.J. MacLachlan, N. Coombs, G.A. Ozin, Periodic mesoporous organosilicas with organic groups inside the channel walls, Nature 402 (1999) 867–871.

[7] S. Guan, S. Inagaki, T. Ohsuna, O. Terasaki, Cubic hybrid organic-inorganic mesoporous crystal with a decaoctahedral shape, J. Am. Chem. Soc. 122 (2000) 5660–5661.

[8] S. Inagaki, S. Guan, T. Ohsuna, O. Terasaki, An ordered mesoporous organosilica hybrid material with a crystal-like wall structure, Nature 416 (2002) 304–307.

[9] A. Comotti, S. Bracco, P. Valsesia, M. Beretta, P. Sozzani, Fast molecular rotor dynamics modulated by guest inclusion in a highly organized nanoporous organosilica, Angew. Chem. Int. Ed. 49 (2010) 1760–1764.

[10] Y. Goto, S. Inagaki, Synthesis of large-pore phenylene-bridged mesoporous organosilica using triblock copolymer surfactant, Chem. Commun. (2002) 2410–2411.

[11] M.P. Kapoor, Q. Yang, S. Inagaki, Self-assembly of biphenylene-bridged hybrid mesoporous solid with molecular-scale periodicity in the pore walls, J. Am. Chem. Soc. 124 (2002) 15176–15177.

[12] M.P. Kapoor, Q. Yang, S. Inagaki, Organization of phenylene-bridged hybrid mesoporous silisesquioxane with a crystal-like pore wall from a precursor with nonlinear symmetry, Chem. Mater. 16 (2004) 1209–1213.

[13] Y. Xia, W. Wang, R. Mokaya, Bifunctional hybrid mesoporous organoaluminosilicates with molecularly ordered ethylene groups, J. Am. Chem. Soc. 127 (2005) 790–798.

[14] N. Mizoshita, Y. Goto, M.P. Kapoor, T. Shimada, T. Tani, S. Inagaki, Fluorescence emission from 2,6-naphthylene-bridged mesoporous organosilicas with an amorphous or crystal-like framework, Chemistry 15 (2009) 219–226.

[15] A. Sayari, W. Wang, Molecularly ordered nanoporous organosilicates prepared with and without surfactants, J. Am. Chem. Soc. 127 (2005) 12194–12195. M. Cornelius, F. Hoffmann, M. Fröba, Periodic mesoporous organosilicas with a bifunctional conjugated organic unit and crystal-like pore walls, Chem. Mater. 17 (2005) 6674–6678.

[16] M. Waki, N. Mizoshita, T. Ohsuna, T. Tani, S. Inagaki, Crystal-like periodic mesoporous organosilica bearing pyridine units within the framework, Chem. Commun. 46 (2010) 8163–8165.

[17] M. Waki, N. Mizoshita, T. Tani, S. Inagaki, Periodic mesoporous organosilicas bearing a high density of metal complexes on pore surfaces, Angew. Chem. Int. Ed. 50 (2011) 11667–11671.

[18] M. Waki, Y. Maegawa, K. Hara, Y. Goto, S. Shirai, Y. Yamada, N. Mizoshita, T. Tani, W.-J. Chun, S. Muratsugu, M. Tada, A. Fukuoka, S. Inagaki, A solid chelating ligand: periodic mesoporous organosilica containing 2,2′-bipyridine within the pore walls, J. Am. Chem. Soc. 136 (2014) 4003–4011.
[19] N. Mizoshita, T. Tani, H. Shinokubo, S. Inagaki, Mesoporous organosilica hybrids consisting of silica-wrapped π-π stacking columns, Angew. Chem. Int. Ed. 51 (2012) 1156–1160.
[20] N. Mizoshita, S. Inagaki, Periodic mesoporous organosilica with molecular-scale ordering self-assembled by hydrogen bonds, Angew. Chem. Int. Ed. 54 (2015) 11999–12003.
[21] S. Inagaki, O. Ohtani, Y. Goto, K. Okamoto, M. Ikai, K. Yamanaka, T. Tani, T. Okada, Light harvesting by a periodic mesoporous organosilica chromophore, Angew. Chem. Int. Ed. 48 (2009) 4042–4046.
[22] H. Takeda, Y. Goto, Y. Maegawa, T. Ohsuna, T. Tani, K. Matsumoto, T. Shimada, S. Inagaki, Visible-light-harvesting periodic mesoporous organosilica, Chem. Commun. (2009) 6032–6034.
[23] H. Takeda, M. Ohashi, T. Tani, O. Ishitani, S. Inagaki, Enhanced photocatalysis of rhenium(1) complex by light-harvesting periodic mesoporous organosilica, Inorg. Chem. 49 (2010) 4554–4559.
[24] Y. Ueda, H. Takeda, T. Yui, K. Koike, Y. Goto, S. Inagaki, O. Ishitani, A visible-light harvesting system for CO_2 reduction using a Ru-II-Re-I photocatalyst adsorbed in mesoporous organosilica, ChemSusChem 8 (2015) 439–442.
[25] H. Takeda, M. Ohashi, Y. Goto, T. Ohsuna, T. Tani, S. Inagaki, A versatile solid photosensitizer: periodic mesoporous organosilicas with ruthenium tris(bipyridine) complexes embedded in the pore walls, Adv. Funct. Mater. 26 (2016) 5068–5077.
[26] N. Mizoshita, Y. Goto, T. Tani, S. Inagaki, Highly fluorescent mesostructured films that consist of oligo(phenylenevinylene)-silica hybrid frameworks, Adv. Funct. Mater. 18 (2008) 3699–3705.
[27] N. Mizoshita, Y. Goto, Y. Maegawa, T. Tani, S. Inagaki, Tetraphenylpyrene-bridged periodic mesostructured organosilica films with efficient visible-light emission, Chem. Mater. 22 (2010) 2548–2554.
[28] N. Tanaka, N. Mizoshita, Y. Maegawa, T. Tani, S. Inagaki, Y.R. Jorapurac, T. Shimada, Synthesis of a spirobifluorene-bridged allylsilane precursor for periodic mesoporous organosilica, Chem. Commun. 47 (2011) 5025–5027.
[29] N. Mizoshita, Y. Goto, T. Tani, S. Inagaki, Efficient visible-light emission from dye-doped mesostructured organosilica, Adv. Mater. 21 (2009) 4798–4801.
[30] N. Mizoshita, M. Ikai, T. Tani, S. Inagaki, Hole-transporting periodic mesostructured organosilica, J. Am. Chem. Soc. 131 (2009) 14225–14227.
[31] M. Ikai, Y. Maegawa, Y. Goto, T. Tani, S. Inagaki, Synthesis of visible-light-absorptive and hole-transporting periodic mesoporous organosilica thin films for organic solar cells, J. Mater. Chem. A 2 (2014) 11857–11865.
[32] Y. Li, F. Auras, F. Löbermann, M. Döblinger, J. Schuster, L. Peter, D. Trauner, T. Bein, A photoactive porphyrin-based periodic mesoporous organosilica thin film, J. Am. Chem. Soc. 135 (2013) 18513–18519.
[33] Y. Maegawa, S. Inagaki, Iridium–bipyridine periodic mesoporous organosilica catalyzed direct C–H borylation using a pinacolborane, Dalton Trans. 44 (2015) 13007–13016.
[34] N. Ishito, H. Kobayashi, K. Nakajima, Y. Maegawa, S. Inagaki, K. Hara, A. Fukuoka, Ruthenium-immobilized periodic mesoporous organosilica: synthesis, characterization, and catalytic application for selective oxidation of alkanes, Chemistry 21 (2015) 15564–15569.
[35] Y. Kuramochi, M. Sekine, K. Kitamura, Y. Maegawa, Y. Goto, S. Shirai, S. Inagaki, H. Ishida, Photocatalytic CO_2 reduction by periodic mesoporous organosilica (PMO)

containing two different ruthenium complexes as photosensitizing and catalytic sites, Chemistry 23 (2017) 10301–10309.
[36] X. Liu, Y. Maegawa, Y. Goto, K. Hara, S. Inagaki, Heterogeneous catalysis of water oxidation by an iridium complex immobilized on bipyridine-periodic mesoporous organosilica, Angew. Chem. Int. Ed. 55 (2016) 7943–7947.
[37] K. Matsui, Y. Maegawa, M. Waki, S. Inagaki, Y. Yamamoto, Transfer hydrogenation of nitrogen heterocycles using recyclable rhodium catalyst immobilized on bipyridin-periodic mesoporous organosilica, Catal. Sci. Technol. 8 (2018) 534–539.
[38] M. Waki, K. Yamanaka, S. Shirai, Y. Maegawa, T. Goto, Y. Yamada, S. Inagaki, Re(bpy)$(CO)_3$Cl immobilized on bipyridine-periodic mesoporous organosilica for photocatalytic CO_2 reduction, Chemistry. 24 (2018) 3846–3853.
[39] S. Ishikawa, Y. Maegawa, M. Waki, S. Inagaki, Immobilization of a molybdenum complex on bipyridine-based periodic mesoporous organosilica and its catalytic activity for epoxidation of olefins, ACS Catal. 8 (2018) 4160–4169.

FURTHER READING

[40] Y. Goto, T. Ohsuna, N. Mizoshita, T. Tani, S. Inagaki, Synthesis of single crystalline anthracene-silica hybrid and its structural and optical properties, Solid State Sci. 13 (2011) 729–735.

CHAPTER THREE

Drug Delivery and Bone Infection

María Vallet-Regí[1], Montserrat Colilla, Isabel Izquierdo-Barba
Department of Chemistry in Pharmaceutical Sciences, School of Pharmacy, Universidad Complutense de Madrid and Networking Research Center on Bioengineering, Biomaterials and Nanomedicine (CIBER-BBN), Madrid, Spain
[1]Corresponding author: e-mail address: vallet@ucm.es

Contents

Abstract

Bone infection represents greatest challenge in public health care with serious social and economic implications. The efforts of the scientific community are focused in the development of innovative and advanced biomaterials with anti-infective properties related to their non-fouling, bactericidal and/or antibiofilm capabilities. This chapter aims at thoroughly surveying the different approaches based on silica mesoporous materials (SMMs) for bone infection management. Bacteria repelling surfaces by *zwitterionization* process, bactericidal effect by implantable devices with antimicrobial local delivery agents and antibiofilm effect by more sophisticated systems based on targeted nanocarriers will be considered.

1. INTRODUCTION

Currently, bone infection remains the greatest challenge in public health care with serious social and economic implications [1]. A major issue to be considered is the increased multi-drug resistant bacteria that is estimated to cause more than 10 million deaths via infection for 2050, and even

The Enzymes, Volume 44
ISSN 1874-6047
https://doi.org/10.1016/bs.enz.2018.08.001
© 2018 Elsevier Inc.
All rights reserved.

overcoming to cancer deaths [2–4]. The combination of these multidrug resistant and bacterial biofilm formation leads to almost unmanageable infections, giving way to a post-antibiotic era, in which conventional antimicrobial treatments are useless. Biofilms are described as complex communities of microorganisms that attach to surfaces and are embedded in a self-produced extracellular matrix [5]. Since these bacteria get increased tolerance against antimicrobial agents and host immune systems, biofilm-associated infectious diseases tend to become chronic [6]. Fig. 1 displays the process of biofilm formation onto a surface.

The first stages of biofilm formation, including the bacteria adhesion and micro-colonies formation, are mainly governed by the electrostatic attraction forces between bacteria and implantable surface. In this step, the electrochemical nature of the biomaterial plays a major role [7,8]. The subsequently stages are mediated by molecular and cellular interactions closely associated with expression of biofilm specific gene clusters with the secretion of protective slime, which makes extremely resistant to both host immune system and antibiotic diffusion [9].

Conventional therapies for infection treatment involve long-term systemic antibiotic administration [10] and surgery [11,12], which have important limitations and significant repercussions for the quality life of the patients. Therefore, there is a great interest to promote the design more effective and non-toxic anti-infective biomaterials as alternative to conventional therapies [13–16].

Recent advances in nanotechnology have offered new prospects to develop novel formulations with anti-infective purposes [17]. These

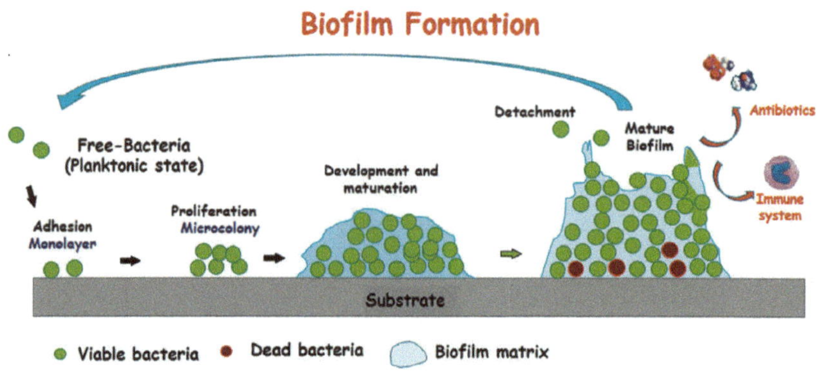

Fig. 1 Biofilm formation since the adhesion of planktonic bacteria onto the substrate surface until the formation of mature biofilm covered by exopolysaccharides film very resistant to the antibiotics and immune system action.

research approaches have been focused in both prevention and treatment of such infections [18,19]. Lack of bacterial adhesion could *per se* involves a reduction of surface infectability [20]. In fact, one of the milestones in this research field consists in the design of implant surfaces able to diminish the bacterial adhesion and subsequently the biofilm formation, being an innovative alternative to manufacturing drug-free surfaces. Moreover, implantable biomaterials with the aim to control pre-existing infections are required to exert a powerful bactericidal action, constituting another milestone in this field [21–24]. They can be employed in the delivering of antimicrobial agents, whose primary scope is the prevention or the treatment through a rigorous control of their release, or by specific-site release through the incorporation of targeting agents [25].

Silica based mesoporous materials (SMMs) are highly attractive candidates for management of bacterial infection due to special features. Some of these characteristics are high specific surface area, large and tunable pore size, drug loading capacity, tuning drug release, simple synthesis, versatility for chemical modification and biocompatibility [21,26–30]. Since they were first proposed for such purposes by Vallet-Regí et al. [31], a very broad variety of concepts and approaches with anti-infective properties have been described. This chapter covers the latest approaches in the design of SMMs with antibacterial purposes, which are summarized in Fig. 2.

First of all, we focus on the *zwitterionization* of SMMs to confer antibacterial adhesion and non-fouling properties. The importance of *zwitterionization* as a cutting-edge technology to provide surfaces preventing the bacterial infection will be reviewed. Second, we study the recent

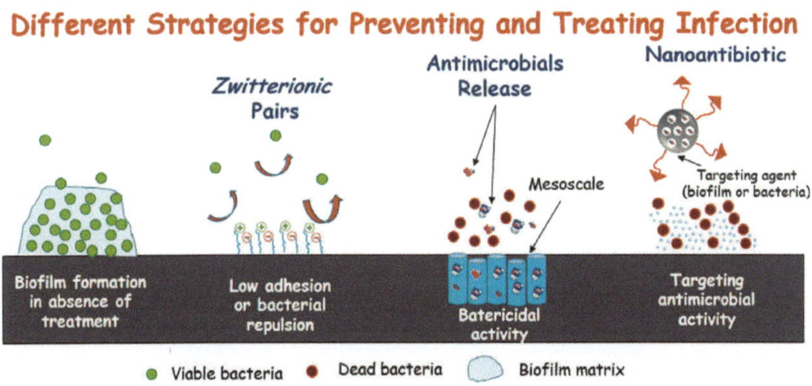

Fig. 2 A look to the different strategies to contrast the establishment of infections revised in this chapter.

advances on these nanostructured biomaterials for local delivery of antimicrobial agents. In this sense, the recent scientific efforts are being addressed to the design of 3D scaffolding based on SMMs that dynamically contribute to the regeneration process, promoting integration into the tissues and avoiding bacterial infection by control release of their cargo. Finally, the development of more sophisticated drug-releasing nanocarriers, based on mesoporous silica nanoparticles (MSNs), with improved capabilities and performances for localized and controlled therapies, is also discussed.

2. PREVENTING BACTERIAL ADHESION: IMPORTANCE OF *ZWITTERIONIZATION*

Over the last few years, different approaches have been described to prevent bacterial contamination of implants. In general, these approaches consist in tuning the surface properties to confer antibacterial adhesion and non-fouling properties, without affecting their biocompatibility. Surface properties, such as topography, chemistry and surface energy become essential pillars to management of bacterial adhesion [18,32]. Thus, several studies have shown that both nanotopography and nanostructure of biomaterial surface play an important role in the control of bacterial adhesion and biofilm formation and could represent a challenging alternative to conventional antibiotics [33–36]. Another important alternative is the chemical modification of biomaterial surface by *zwitterionization* process, which has raised as a groundbreaking strategy to confer surfaces of high resistance to nonspecific protein adsorption, bacterial adhesion and/or biofilm formation [37,38]. This section focuses on the general features and potential applications of *zwitterionization* process on SMMs to prevent the bacterial infection, keeping of biocompatible properties.

2.1 Definition and Characteristics of *Zwitterionization* Process

Zwitterionic surface possesses an equal number of both positively and negatively charged groups maintaining overall electrical neutrality (Fig. 3). The non-fouling capability of *zwitterionic* surfaces is related to the formation of a hydration layer near the surface, forming a physical and energetic barrier that hampers nonspecific proteins adhesion becoming an important part in affording interfacial bioadhesion resistance [39]. The *zwitterionization* process consists in the functionalization of surface at atomic level. Initially, the research efforts were addressed on functionalizing with *zwitterionic* polymers bearing mixed positively and negatively charged moieties within the same

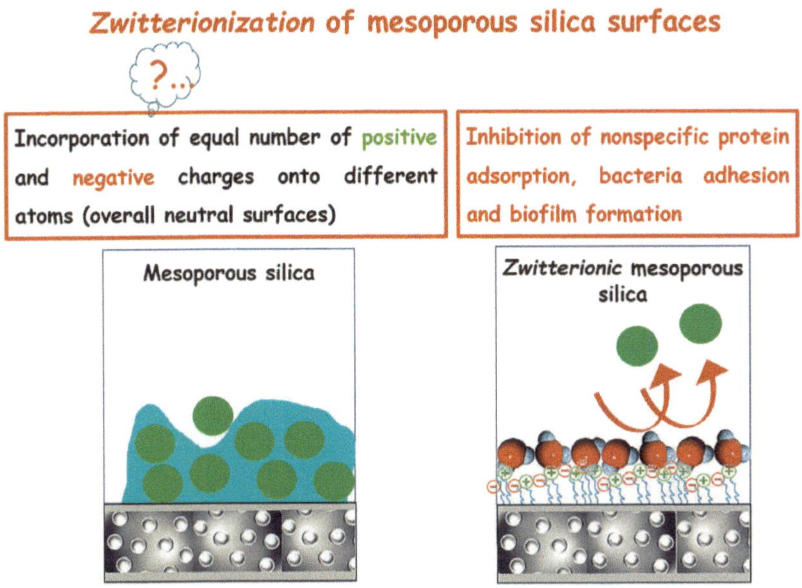

Fig. 3 Definition and properties of *zwitterionization* of mesoporous silica materials.

chain and overall charge neutrality [40–42]. Moreover, it is possible to confer *zwitterionic* surfaces by grafting low-molecular weight moieties bearing the same number of negative and positive charges as amino acids [43–45], sulfobetaine derivatives [46–48] or dopamine [49–51], which exhibit *zwitterionic* nature as function of the pH. Although the reported methods offer certain advantages over *zwitterionic* polymers in terms of biocompatibility, the procedures involve different steps with several intermediate products, which make complex the process.

Another strategy consists on the use of more direct and simple grafting methods based on sol-gel chemistry using organosilanes [52]. This strategy has been applied to SMMs, taking advantage of the presence of high amount of hydroxyl (–OH) groups on their surface and their facility to be covalently attaching simultaneously with two organosilanes exhibiting functional groups with positive and negative charges, respectively. One of the main advantages of this methodology is the possibility to tailor the *zwitterionic* properties by adjusting the molar ratio of the two organosilanes. Thus, it was in 2010 when Colilla et al. described, for the first time, the preparation of *zwitterionic* mesoporous SBA-15 material containing $-NH_3^{\oplus}/-COO^{\ominus}$ *zwitterionic* pairs by co-condensation of two alkoxysilanes, aminopropyltrimethoxysilane (APTES) and carboxyethyl silanetriol sodium salt (CES), respectively [52]. This *zwitterionic* material is able to inhibit unspecific protein adsorption and bacterial adhesion [53]. Fig. 4 shows as representative example concerning the design of *zwitterionic* SBA-15 by using an alkoxysilane containing primary and secondary amine groups (*N*-(2-aminoethyl)-3-aminopropyl-trimethoxysilane) (DAMO) [54]. As it can be observed in Fig. 4, its *zwitterionic* nature is provided by the presence of $-NH_3^{\oplus}/-SiO^{\ominus}$ and $>NH_2^{\oplus}/-SiO^{\ominus}$ *zwitterionic* pairs. Microbiological assays regarding bacterial adhesion using *Staphylococcus aureus* displayed a noticeable inhibition of 99.9% with respect to pristine SBA-15. The coexistence of $-NH_3^{\oplus}/SiO^{\ominus}$ and $-NH_2^{\oplus}/SiO^{\ominus}$ pairs onto Zwitter-SBA15 avails its bacterial-repelling properties.

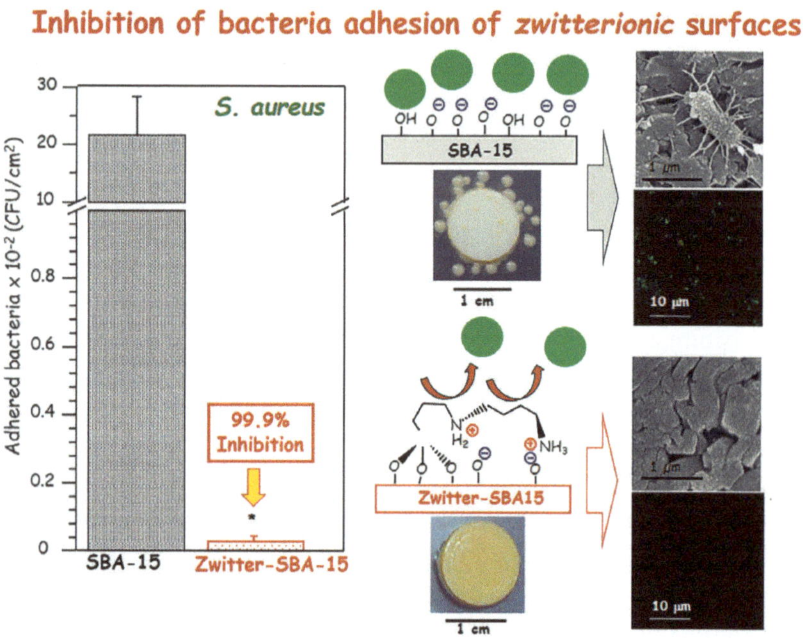

Fig. 4 Schematic representation of pure silica SBA-15 and *zwitterionic* SBA-15 nanostructured materials. Zwitter-SBA15 was synthetized by co-condensation route in the presence of DAMO. Histogram showing bacteria adhesion assay by counting of colony forming units (CFU) of *S. aureus* after 90 min of culture onto SBA-15 and Zwitter-SBA15 surfaces. Statistical significance: $*P < 0.01$. SEM and confocal microscopy images of this study, evidencing the absence of bacteria in Zwitter-SBA15 sample. For confocal microscopy purposes the samples were stained with Baclight® Kit™.

2.2 Biocompatibility of *Zwitterionic* Mesoporous Silica Materials

In terms of biocompatibility, the *zwitterionization* process must besides to prevent bacterial adhesion allow adequate host cell colonization [18]. Fig. 5 shows the process of biomaterial surface colonization through the concept of "Race of the surface" with three main actors implicated; bacteria, proteins and host cells (osteoblasts). In the first steps, mediated by the biomaterial surface properties, the non-specific protein adhesion constitutes a key factor [55]. There is "window opportunity" between bacteria and osteoblast. If the race for the surface is won by host cells, which achieve irreversible attachments on the biomaterial surface, the presence of a continual cell layer makes it complicated for bacteria to begin biofilm formation [56]. On the contrary, if the bacteria wins, the process is irreversible through the biofilm formation [57]. *In vitro* studies using osteoblastic-like cells (Fig. 5 down) revealed that zwitterionic surfaces are able to appropriately adhere, colonize, and spread of osteoblast [38]. This different behavior between bacteria and osteoblast cells could be due mainly to two main reasons. First, the stiffness and small size of the bacteria compared with host cells (bacteria size (*ca.* 1 μm)

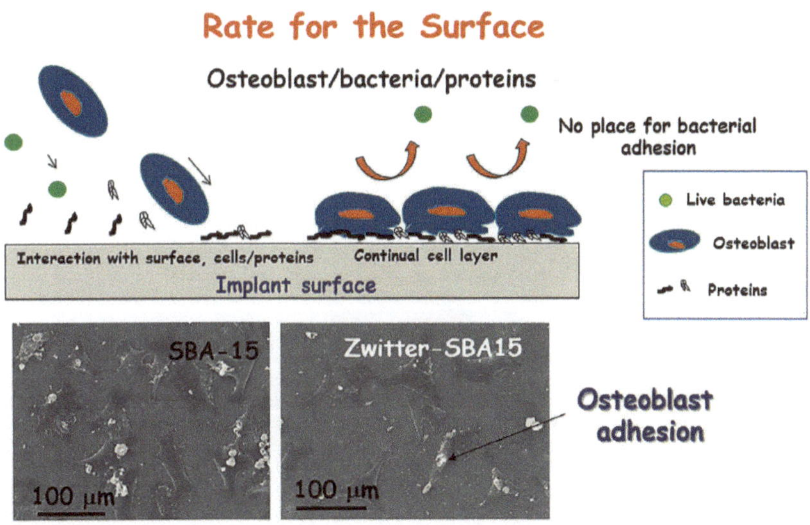

Fig. 5 Theory of "race for the surface" in the behavior of implants able to be colonized by host cells (osteoblast) at the same time that inhibit the bacteria adhesion and biofilm formation. SEM micrographs showing the well-osteoblast colonization onto the surface of SBA-15 and Zwitter-SBA15, respectively.

and eukaryotic cell size (*ca.* 50 μm)), could be responsible of their capacity to distinguish variations in the biomaterials surfaces at the nanoscale. Finally, the adhesion mechanism of bacteria is mainly driven by electrostatic attractive forces, in which the electrochemistry of the biomaterial surface is a pivotal factor [7,8].

2.3 Dual Functionality in *Zwitterionic* Mesoporous Materials

Taking advantage of the intrinsic characteristics of SMMs, it is possible to host antimicrobial agents and release them in a controlled manner. Therefore, it is possible to provide them dual functionality by combination of bacterial repellent and killing bacteria capabilities [29,30,58–60]. Fig. 6 displays as Zwitter-SBA15 is capable to load cephalexin, used as model antibiotic, with a value of *ca.* 12 mg g^{-1} into the mesoporous structure and release it in a sustained fashion for more than 15 days. This dual functionality, i.e., antibacterial adhesion and antibiotic drug delivery, constitutes a step forward for development of advanced bioceramics for the management of bone implant infection.

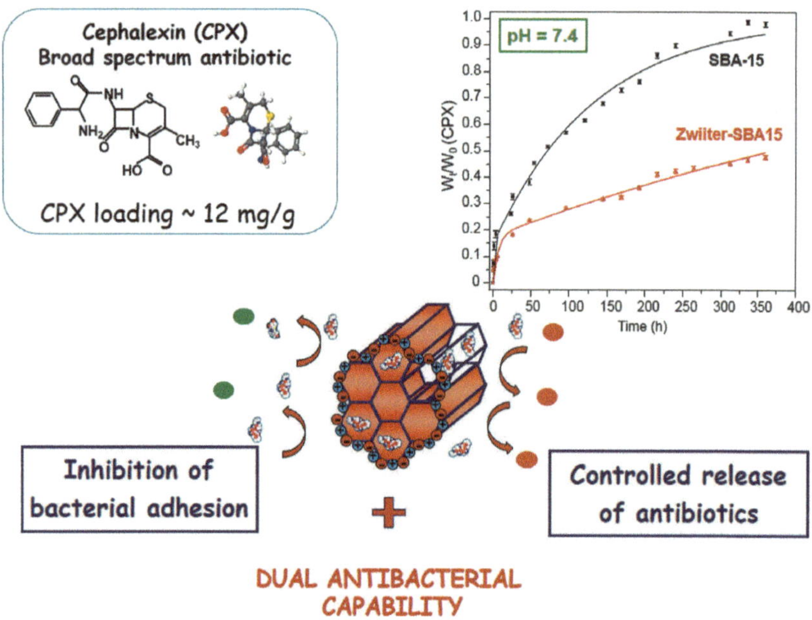

Fig. 6 Dual functionality with antibacterial purposes of SMMs by: (i) zwitterionization of their surfaces inhibiting the bacterial adhesion and (ii) load and release capability of antimicrobial agent in the mesoporous arrangement.

3. MESOPOROUS MATERIALS IN THE TREATMENT OF INFECTION

As it has been commented in Section 1, SMMs have been riding high by the biomedical scientific community since they were described as unique matrices for drug delivery [31]. Their exceptional textural and structural properties, as controllable pore structure, large pore volume and surface area, easily functionalizable, tuneable pore diameter and morphology, biocompatibility and tuneable composition make them excellent candidates as drug delivery devices in different biomedical application, including the management of infection [21,61–64]. In 2006, SMMs were purposed as third-generation bioceramics in the context of drug delivery system and bone tissue regeneration thank to their surfaces features as biocompatibility and bioactivity, and their capability to load and controllable release different drugs for different pathologies [65,66]. Fig. 7 shows the structural and textural properties of SMMs and their implications in the load and release of different drugs. In this context different antibiotics and antimicrobial peptides have been loaded and released since these matrices [67–71]. The most widely employed SMMs for such applications have been MCM-41, SBA-15 and MCM-48. Important challenges have been achieved in this field as: (i) controllable antibiotic release as function of textural, chemical and structural characteristics of the matrices, (ii) sustained and prolonged time release of wide spectrum antibiotics and pH-dependent release [72], (iii) stability of antimicrobial agents into mesoporous structure, which are able to keep the biological activity of the antibiotic once they are released from these matrices [70,73] and (iv) possibility to introduce different antimicrobial agents into

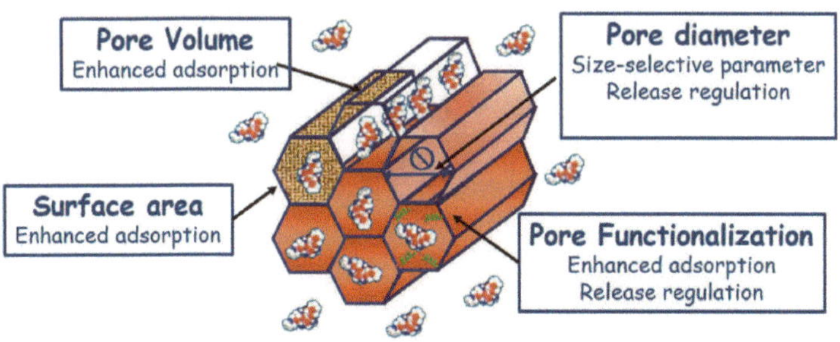

Fig. 7 Parameters governing the loading and release rate of biologically active molecules into SMMs.

mesoporous structure, which lead a design of useful combined therapy in the infection treatment [74].

SMMs can be fabricated in a wide range of morphologies, bulk, fibers, films, rods, monoliths, macroporous scaffolds, spheres, etc. In this regard, the scientific efforts are addressed to custom-made design depending on the biomedical application. Among them, in this chapter we tackle the design of both 3D scaffolds and mesoporous silica nanoparticles (MNSs) for the treatment of the infection, which represents the cutting edge in customized therapies. In this sense, MSNs, with sizes in the range of 20–250 nm, have found their niche in the field of nanomedicine, specifically in the systemic treatment of complex diseases as cancer. However, although very recent, they begin walking in the infection treatment with promising perspectives.

3.1 Design of Multifunctional 3D Scaffolds in Infection Treatment

Multifuctional 3D scaffolds based on SMMs constitute a powerful and promising tool in implantable devices for bone tissue regeneration applications [66,75,76]. The Holy Grail in this field is to design these 3D scaffolds capable to inhibit and/or abolish bacterial infection meanwhile allowing osteoblast cells colonization. These advanced multifunctional scaffolds would constitute an added value, establishing a great technological advance to achieve better clinical outcomes. There are many efforts that have been addressed for the fabrication of 3D scaffolds that dynamically contribute to the regeneration process, stimulating the integration, osteoconduction and angiogenesis and evading bacterial infection [22,77,78]. In this sense, the local administration of antimicrobial agents could minimize the feared side effects improving the bioavailability of the antimicrobial and easily reaching the therapeutic effective dose in the target site. Furthermore, the added value with their capability to osteointegration once the infection in avoid marks significant advantages over conventional therapies.

For infection treatment, it is important the combination of different antimicrobial agents with sustained and controlable release over time to avoid the bacterial resistance [79]. In this regards, an initial fast release of an antibiofilm drug to be able to destroy the biofilm capsule and subsequently more sustained and prolonged release of the different wide-spectrum antibiotics would be desired [18]. Recently, 3D multidrug scaffolds, containing rifampin (RIF), levofloxacin (LEVO) and vancomycin (VAN), have been designed by rapid prototyping technique [78]. These 3D scaffolds are composed by mixture of a *nanocomposite* mesostructured bioceramic and polyvinyl alcohol (PVA), forming a slurry and posterior conformed by robocasting technique.

These 3D scaffolds, exhibiting hierarchical macro-mesoporous structure formed by different range of porosity, were external coated of gelatin-glutaraldehyde layer (GEL). The different antimicrobial agents have been localized in different compartments to achieve different release kinetics. Thus, the LEVO was localized into mesoporous structure, the VAN in the PVA polymeric framework and RIF in the external GEL layer. As you can see in Fig. 8, the release profiles of different drugs show an early and fast release of RIF followed by sustained and prolonged release of VAN and LEVO. This sequential release is able to destroy *Staphylococcus* biofilm as it can be observed in Fig. 9.

In vitro microbiological assays have been carried out to determine their effectiveness against *S. aureus* biofilms. These antimicrobial studies have been performed for scaffolds containing only LEVO and VAN ($MBG_{LEVO}PVA_{VAN}$)

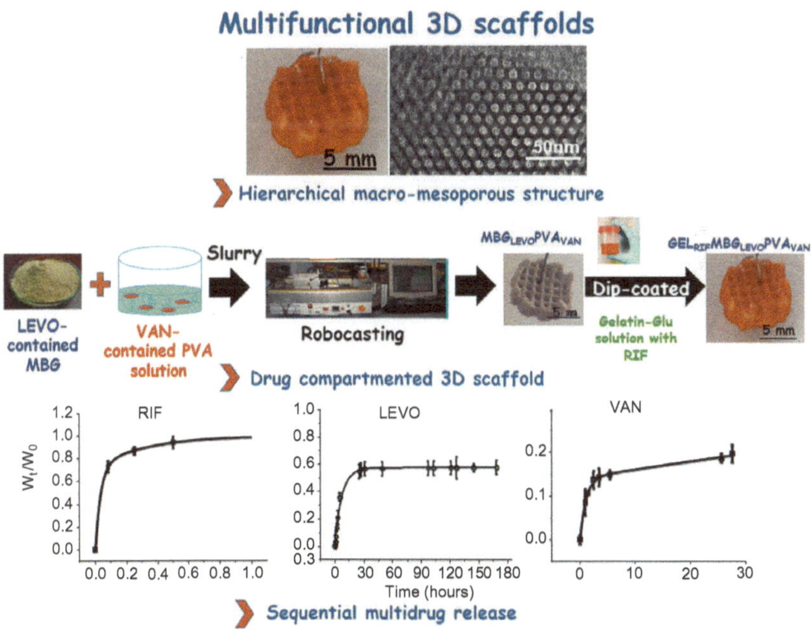

Fig. 8 Therapeutic multidrug 3D scaffolds containing rifampin (RIF), levofloxacin (LEVO) and vancomycin (VAN). Rapid prototyping fabrication of multifunctional scaffolds formed by a mixture of MBG and PVA ($MBG_{LEVO}PVA_{VAN}$) and posterior coating with Gel-Glu layer containing RIF ($GEL_{RIF}MBG_{LEVO}PVA_{VAN}$). Release profiles of different antibiotics from these multifunctional 3D scaffolds based on sequential antibiotic release, early and fast release of RIF and subsequently sustianed release of both VAN and LEVO.

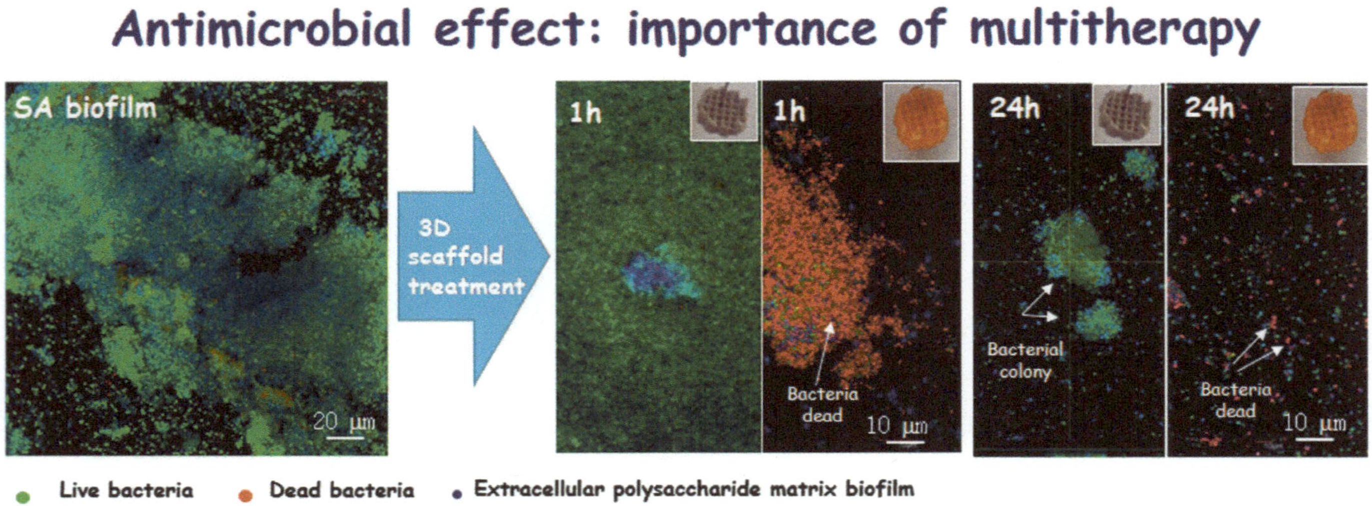

Fig. 9 Antimicrobial effect against *S. aureus* (SA) biofilm of both 3D scaffolds $GEL_{RIF}MBG_{LEVO}PVA_{VAN}$ and $MBG_{LEVO}PVA_{VAN}$, evidencing the importance of multitherapy. After treatment with $MBG_{LEVO}PVA_{VAN}$ (containing only VAN and LEVO) still appear small fragments of biofilm of size of 20–30 µm after 24 h. However, with $GEL_{RIF}MBG_{LEVO}PVA_{VAN}$ scaffolds, it is evident the rapid antimicrobial effect, showing the total destruction of biofilm.

and complete scaffolds containing also RIF ($GEL_{RIF}MBG_{LEVO}PVA_{VAN}$) 3D scaffolds. The initial biofilm showed a typical structure formed by colony live bacteria (green) covered by protective mucopolysaccharide matrix (blue). After treatment with the different scaffolds, significant differences can be observed between both scaffolds, determining the importance of combined therapy. While $MBG_{LEVO}PVA_{VAN}$ containing both LEVO and VAN is not able to destroy completely the biofilm, observing large amount of live bacteria forming colonies with the protective layer, in the case of $GEL_{RIF}MBG_{LEVO}PVA_{VAN}$ a complete destruction of biofilm is observed. These findings agree with *in vitro* drug release (Fig. 8), where a sudden release of antibiofilm (RIF) provokes a greater vulnerability of the biofilm for the subsequently release of VAN and LEVO, making this therapeutic multidrug scaffolds more effective.

To achieve an *in vitro* test more representative to *in vivo* conditions, *in vitro* bacteria-osteoblast co-culture model has been performed. Both MC3T3-E1 osteoblast-like cells and *S. aureus* were seeded on the scaffold surfaces containing or not antimicrobial agents (Fig. 10). For free-antimicrobial scaffolds, the confocal microscopies show the survival of the bacteria, which begin to invade the osteoblasts in the first hours of incubation. After 24 h of testing, apoptotic cells appear as rounded red spots while the biofilm maturation keeps in successful progress until a fully developed state with osteoblast complete disruption and disappearance. On the contrary, after treatment with antimicrobial-containing scaffolds is observed the totally absence of bacterial population adhered to the material surface and osteoblasts healthily attached and growing. Although further in vivo research is necessary, all these obtained results reveal the potential of these 3D multidrug scaffolds to be used as therapeutic platforms for treatment and prevention of bone infection.

3.2 Mesoporous Silica Nanoparticles for Infection Treatment

The design of nanodevices that selectively transport antimicrobial agents to the target site and, once there, delivery them in a controlled fashion constitutes one of the most important challenges in infection therapy [80–85]. Bacteria or biofilm-targeted nanoparticles offer promising approaches to enhance drug efficacy compliance and safety [86,87]. Different nanocarriers from liposomal to polymer based nano-drug carriers, metallic vectors, as gold nanoparticles, have been proposed for these aims due to their nontoxic nature and effectiveness [88,89]. Among nanocarriers, mesoporous silica

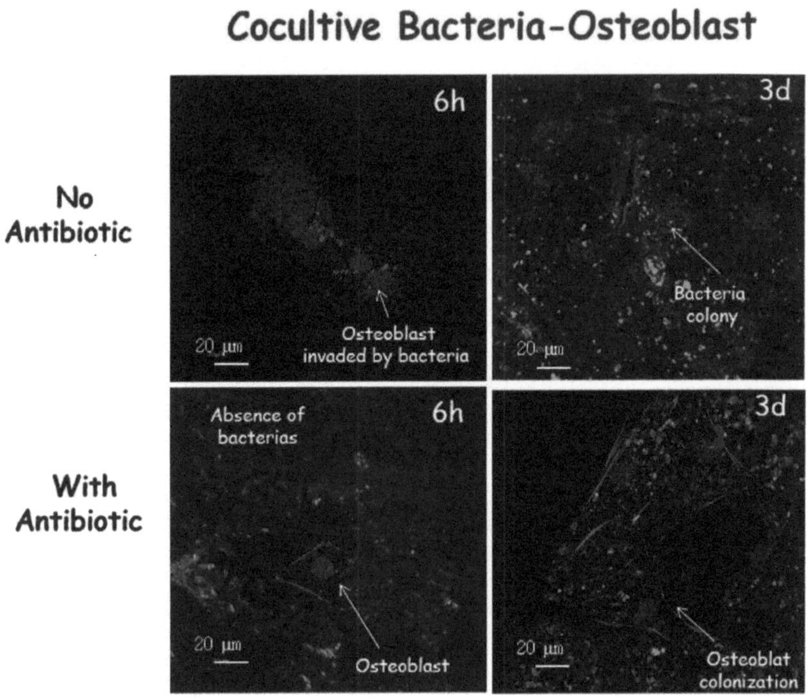

Fig. 10 *In vitro* competitive co-culture MC3T3-E1 preosteoblast and *S. aureus* after 6 h, and 3 days of incubation onto 3D scaffolds in absence and presence of antibiotic load, respectively. Material refraction in green, MC3T3-E1 nuclei and bacteria in blue (DAPI) and actin-fibrous of preosteoblast cytoplasm in red (phalloidin).

nanoparticles (MSNs) are excellent candidates to develop targeted drug delivery devices due to their ability to introduce a wide range of antimicrobial agents (contained within their mesoporous structure), and their ability to release in a controlled fashion and in the target, thanks to his high versatility in terms of chemical modification [26,90–92]. Furthermore, it has been demonstrated widely their biocompatibility both *in vitro* and *in vivo* [93–95] and stability in different biological media [27,96].

Concerning the biofilm-targeted nanodevices, recent studies have demonstrated that control of surface functionality on the dense nanoparticles surface provides an effective approach to predict the nanoparticle behavior in the biofilm. As you can see in Fig. 11 nanoparticles functionalized with different charged moieties exhibit different behavior in the rate of penetrability and distribution into the bacteria biofilm [97]. Neutral and anionic nanoparticles cannot internalize into the biofilm, while cationic nanoparticles can easily penetrate into biofilm. With these cationic

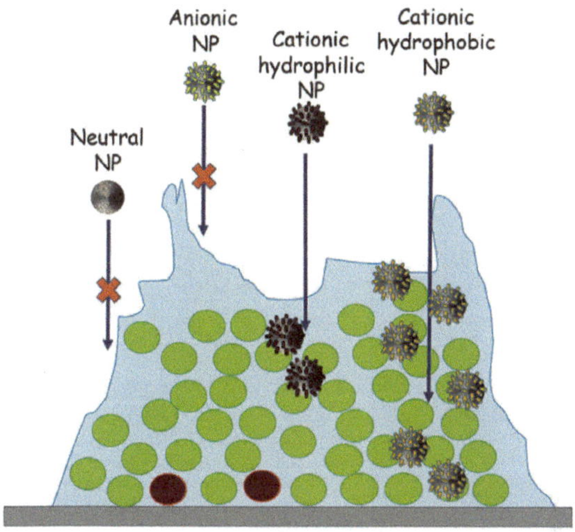

Fig. 11 Schematic illustration of surface functionality-controlled nanoparticles penetration into biofilms.

nanoparticles, it has been determined that the penetrability rate is higher in hydrophobic nanoparticles compared with hydrophilic ones.

These concepts have been applied to MSNs in order to internalize into bacterial biofilm but with the added value that once the nanosystem has reached the biofilm the anti-bacterial activity is determined by the physicochemical properties of the entrapped antimicrobials. Thus it has been designed a "nanoantibiotic" formed by MSNs functionalized with *N*-(2-aminoethyl)-3-aminopropyltrimethoxysilane (DAMO), which deploys positively charged amine groups on the MSN surface and load with LEVO (fluoroquinolone antibiotic) into the mesoporous structure (Fig. 12A) [98]. The possible mechanism of this new "nanoantibiotic" is represented in Fig. 12B, where the engineered nanodevices and charged positively are able to penetrate into the biofilm and once there, release the LEVO loaded, exerting their antimicrobial effect.

To directly observing the internalization capability into biofilm confocal microscope studies were carried out. Fig. 13 (left) displays more representative results showing the mucopolysaccharide layer of biofilm stained in blue and the green-labeled nanoparticles. To comparative proposes these studies were also performed on bare MSNs. The obtained results evidence significant differences between MSN and MSN-DAMO. While for pristine MSNs, the nanoparticles are localized in the vicinity of the biofilm, in MSN-DAMO the nanoparticles are internalized within biofilm,

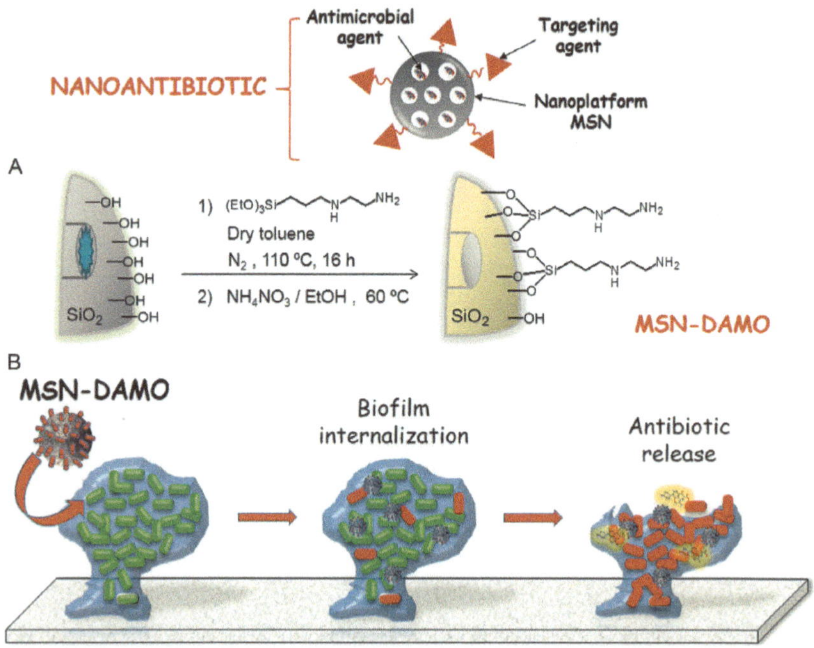

Fig. 12 Design of "nanoantibiotic" consisting of MSN platform loaded with antimicrobial agent (LEVO) and grafted in their outermost surface with DAMO, which acts as biofilm targeting agent. (A) Synthesis scheme and (B) mechanism of action of this "nanoantibiotic."

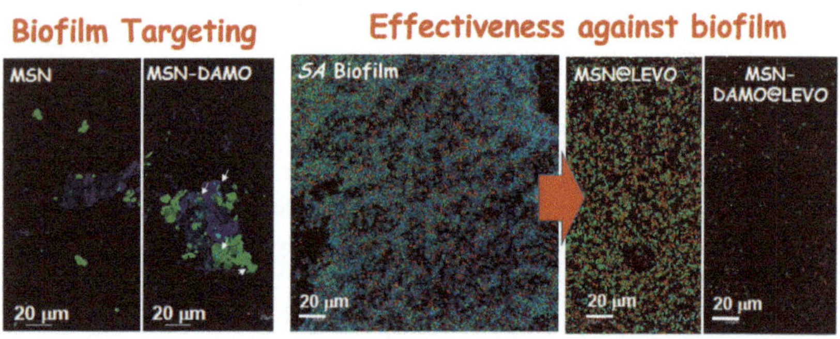

Fig. 13 *Left*: Evaluation of biofilm internalization. Confocal microscopy showing the targeting effect of targeting biofilm effect of MSN-DAMO. *Right*: Confocal microscopy study of the antimicrobial activity of the LEVO loaded MSNs materials onto *S. aureus* biofilm. The confocal images show the biofilm preformed onto covered glass-disk before and after 90 min of incubation with MSN@LEVO and MSN-DAMO@LEVO. Live bacteria are stained in green, dead bacteria in red and the protective biolayer, i.e., the extracellular polysaccharide matrix biofilm in blue.

demonstrating the capability of this amine-functionalized nanosystem to penetrate into the mucopolysaccharide matrix. Once demonstrated their internalization capability of free-antibiotic nanosystems, *in vitro* antimicrobial efficacy against preformed biofilm was carried out with the complete "nanoantibiotic," i.e., MSN-DAMO@LEVO. Fig. 13 (right) summarized these studies. Initially, the preformed *S. aureus* biofilm displays a typical structure composed of colony live bacteria (green) covered by a protective mucopolysaccharide matrix (blue). After 90 min of incubation with MSN@LEVO and MSN-DAMO@LEVO, notable differences are observed. MSN@LEVO is not able to destroy the biofilm, appearing large amount live bacteria forming dense mantle and small amounts of scattered dead bacteria due to the action of the LEVO released from this material. On the contrary, MSN-DAMO@LEVO triggers the biofilm total disruption appearing only few green scattering almost negligible. Such differences are consequence of the biofilm penetrability of these engineered nanoparticles.

On the other hand, bacteria-targeted nanoparticles offer promising expectations to improve drug efficacy to avoid the antimicrobial resistance. Both Gram-positive and Gram-negative bacteria have different components in their outer membranes, which are deprotonated to produce negative charges [99]. Cationic polymers and moieties have been described as bacterial detection probes through electrostatic interaction [100] and they are highly toxic and widely used in the bacteria immobilization and detection [101,102]. Dendrimers are single molecular weight, tree-like globular macromolecules with structural uniformity consisting of a central core, branching layers, and numerous end groups [103]. Their intrinsic features make them good nanoplatforms in biomedicine areas, as protein mimics, drug delivery systems, imaging and antiviral and antibacterial agents [104,105]. In the case of polycationic dendrimers, they exhibit high concentration of positively charged amine groups on the surface, able to bind to negatively charged bacteria cell walls, increase bacterial membrane permeability and internalize inside the bacteria [106,107]. With the aim to design a new "nanoantibiotic" with specific bacteria-targeting, MSNs were functionalized in the external surface with the polypropyleneimine (PPI) dendrimer of third generation (G3) [108] and loaded also with LEVO inside of mesoporous structure as antimicrobial agent. A schematic representation of this synthesis is displayed in Fig. 14 (top). In this case, the positive charge of this "nanoantibiotic" should allow the penetration into the

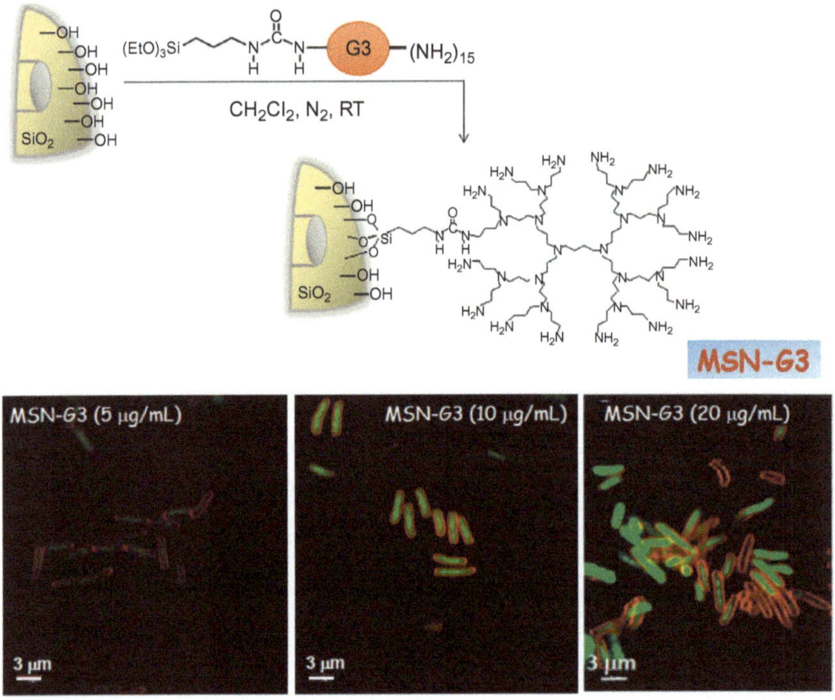

Fig. 14 Synthetic route used for the external functionalization of MSNs with dendrimer G3 as bacteria targeting agent. Evaluation of bacteria internalization as function of nanoparticles concentration.

negatively charged bacteria wall and once there the LEVO loaded should exert its antimicrobial effect [109]. Fig. 14 (bottom) shows the success in penetrability of green-labeled MSN-G3 into Gram-negative bacteria. Moreover, this internalization rate is dose-dependent, due to when the concentration of nanoparticles is increased the amount of nanoparticles also is increased as we can observe in Fig. 14. In order to determine the bacteria internalization mechanism, different positively charged functionalized MSNs have been studied and compared. Then, bacteria internalization studies have been performed on both MSN-DAMO and MNS-G3, showing only internalization in this last case. Difference in the successful internalization of both hybrid materials may be attributed to the G3 dendritic skeleton, which provides MSN-G3 of high surface flexibility with more interaction points compared with MSN-DAMO. Therefore, MSN-G3 not only binds to the negatively charged phospholipid membranes but also provokes a slight change in membrane permeability, which finally leads to nanosystem internalization into the bacteria, as it has been reported elsewhere for other systems [110].

4. CONCLUDING REMARKS

The loss of efficacy of conventional antimicrobial therapies represents an important challenge and urge to find new preventive or treatment strategies able of counteracting this issue. Silica mesoporous materials (SMMs) constitute a powerful alternative to these conventional therapies in the management of bone infection. Since they were first proposed as excellent matrices for drug delivery by Vallet-Regí in 2001 a very broad diversity of concepts and approaches with anti-infective properties have been described. All of them have been addressed to custom-made therapies more effective and safe to avoid the bacterial resistance and biofilm. Bacterial-repelling surfaces by *zwitterionization* of mesoporous materials represent an advanced and promising technology for the design of biocompatible surfaces to prevent the bacterial infection. The recent advances of SMMs, as local delivery systems of antimicrobials, are being focused to the fabrication of modern implantable devices based of 3D scaffolding that dynamically promote the regeneration process, allowing their integration into the tissues and evading bacterial infection by control release of their cargos. Furthermore, bacteria or biofilm-targeted nanoparticles offer real promising approaches to enhance drug efficacy compliance and safety. Given their vast therapeutic potential of SMMs with anti-infective properties, much research effort is being committed to be able to translate these significant advances from bench to bedside.

ACKNOWLEDGMENTS

This work was supported by European Research Council ERC-ADG-2015-VERDI Project ID: 694160; Ministerio de Economía y Competitividad (MINECO) Grant MAT2015-64831-R.

REFERENCES

[1] G. Taubes, The bacteria fight back, Science 321 (2008) 356–361.
[2] World Health Organization, Antimicrobial Resistance: Global Report on Surveillance 2014, WHO, 2014.
[3] M.E.A. de Kraker, A.J. Stewardson, S. Harbarth, Will 10 million people die a year due to antimicrobial resistance by 2050? PLoS Med. 13 (2016) e1002184.
[4] B.D. Lushniak, Antibiotic resistance: a public health crisis, Public Health Rep. 129 (2014) 314–316.
[5] C.R. Arciola, D. Campoccia, L. Montanaro, Implant infections: adhesion, biofilm formation and immune evasion, Nat. Rev. Microbiol. 16 (2018) 397–409.
[6] P.S. Stewart, J.W. Costerton, Antibiotic resistance of bacteria in biofilms, Lancet 358 (2001) 135–138.

[7] L. Ploux, A. Ponche, K. Anselme, Bacteria/material interfaces: role of the material and cell wall properties, J. Adhes. Sci. Technol. 24 (2010) 2165–2201.
[8] J. Bohmler, A. Ponche, K. Anselme, L. Ploux, Self-assembled molecular platforms for bacteria/material biointerface studies: importance to control functional group accessibility, ACS Appl. Mater. Interfaces 5 (2013) 10478–10488.
[9] G. Laverty, S.P. Gorman, B.F. Gilmore, Biomolecular mechanisms of staphylococcal biofilm formation, Future Microbiol. 8 (2013) 509–524.
[10] J.T. Mader, M.E. Shirtliff, S.C. Bergquist, J. Calhoun, Antimicrobial treatment of chronic osteomyelitis, Clin. Orthop. Relat. Res. 360 (1999) 47–65.
[11] R.O. Darouiche, Treatment of infections associated with surgical implants, N. Engl. J. Med. 350 (2004) 1422–1429.
[12] B. Parsons, E. Strauss, Surgical management of chronic osteomyelitis, Am. J. Surg. 188 (2004) 57–66.
[13] J.H. Lee, S.H. Jeong, S. Cha, S.H. Lee, A lack of drugs for antibiotic-resistant Gram-negative bacteria, Nat. Rev. Drug Discov. 6 (2007) 29–40.
[14] H. Le Moual, S. Gruenheid, Resistance to antimicrobial peptides in Gram negative bacteria, FEMS Microbiol. Lett. 330 (2012) 81–89.
[15] K. Dong, E. Ju, N. Gao, Z. Wang, J. Ren, X. Qu, Synergistic eradication of antibiotic-resistant bacteria based biofilms in vivo using a NIR-sensitivenanoplatform, Chem. Commun. 52 (2016) 5312–5315.
[16] D. Campoccia, L. Montanaro, C.R. Arciola, The significance of infection related to orthopedic devices and issues of antibiotic resistance, Biomaterials 27 (2006) 2331–2339.
[17] S.J. Lam, N.M. O'Brien-Simpson, N. Pantarat, A. Sulistio, E.H.H. Wong, Y.-Y. Chen, J.C. Lenzo, J.A. Holden, A. Blencowe, E.C. Reynolds, G.G. Qiao, Combating multidrug-resistant Gram-negative bacteria with structurally nanoengineered antimicrobial peptide polymers, Nat. Microbiol. 1 (2016) 16162.
[18] D. Campoccia, L. Montanaro, C.R. Arciola, A review of the biomaterials technologies for infection-resistant surfaces, Biomaterials 34 (2013) 8533–8554.
[19] S. Mei, H. Wang, W. Wang, L. Tong, H. Pan, C. Ruan, Q. Ma, M. Liu, H. Yang, L. Zhang, Y. Cheng, Y. Zhang, L. Zhao, P.K. Chu, Antibacterial effects and biocompatibility of titanium surfaces with graded silver incorporation in titania nanotubes, Biomaterials 35 (2014) 4255–4265.
[20] J. Hasan, R.J. Crawford, E.P. Ivanova, Antibacterial surfaces: the quest for a new generation of biomaterials, Trends Biotechnol. 31 (2013) 295–304.
[21] M. Vallet-Regí, I. Izquierdo-Barba, M. Colilla, Structure and functionalization of mesoporous bioceramics for bone tissue regeneration and local drug delivery, Philos. Trans. R. Soc. A 370 (2012) 1400–1421.
[22] T. Dvir, B.P. Timko, D.S. Kohane, R. Langer, Nanotechnological strategies for engineering complex tissues, Nat. Nanotechnol. 6 (2011) 13–22.
[23] V. Guarino, M.G. Raucci, A. Ronca, V. Cirillo, L. Ambrosio, Multifunctional scaffolds for bone regeneration. Part II: biomaterial substitute scaffolds and implants for bone repair, in: K. Mallick (Ed.), Bone Substitute Biomaterials, Elsevier Ltd, 2014, ISBN: 978-0-85709-497-1, pp. 95–117.
[24] C. Wu, Y. Zhou, M. Xu, P. Han, L. Chen, J. Chang, Y. Xiao, Copper-containing mesoporous bioactive glass scaffolds with multifunctional properties of angiogenesis capacity, osteostimulation and antibacterial activity, Biomaterials 34 (2013) 422–433.
[25] H. Koo, R.N. Allan, R.P. Howlin, P. Stoodley, L. Hall-Stoodley, Targeting microbial biofilms: current and prospective therapeutic strategies, Nat. Rev. Microbiol. 15 (2017) 740–755.
[26] M. Vallet-Regi, M. Colilla, I. Izquierdo-Barba, M. Manzano, Mesoporous silica nanoparticles for drug delivery: current insights. Molecules 23 (2017) 47, https://doi.org/10.3390/molecules23010047.

[27] J.L. Paris, M. Colilla, I. Izquierdo-Barba, M. Manzano, M. Vallet-Regi, Tuning mesoporous silica dissolution in physiological environments: a review, J. Mater. Sci. 52 (2017) 8761–8771.
[28] M. Vallet-Regí, Nanostructured mesoporous silica matrices in nanomedicine, J. Intern. Med. 267 (2010) 22–43.
[29] M. Vallet-Regí, Ordered mesoporous materials in the context of drug delivery systems and bone tissue engineering, Chem. Eur. J. 12 (2006) 5934–5943.
[30] M. Vallet-Regí, F. Balas, D. Arcos, Mesoporous materials for drug delivery, Angew. Chem. Int. Ed. 46 (2007) 7548–7558.
[31] M. Vallet-Regí, A. Rámila, R.P. del Real, J. Pérez-Pariente, A new property of MCM-41: drug delivery system, Chem. Mater. 13 (2001) 308–311.
[32] K. Anselme, P. Davidson, A.M. Popa, M. Giazzon, M. Liley, L. Ploux, The interaction of cells and bacteria with surfaces structured at the nanometre scale, Acta Biomater. 6 (2010) 3824–3846.
[33] C. Díaz, P.L. Schilardi, R.C. Salvarezza, M.F.L. de Mele, Nano/microscale order affects the early stages of biofilm formation on metal surfaces, Langmuir 23 (2007) 11206–11210.
[34] A. Tiraferri, C.D. Vecitis, M. Elimelech, Covalent binding of single-walled carbon nanotubes to polyamide membranes for antimicrobial surface properties, ACS Appl. Mater. Interfaces 3 (2011) 2869–2877.
[35] I. Izquierdo-Barba, J.M. García-Martín, R. Álvarez, A. Palmero, J. Esteban, C. Pérez-Jorge, D. Arcos, M. Vallet-Regí, Nanocolumnar coatings with selective behavior towards osteoblast and Staphylococcus aureus proliferation, Acta Biomater. 15 (2015) 20–28.
[36] E.P. Ivanova, J. Hasan, H.K. Webb, V.K. Truong, G.S. Watson, J.A. Watson, V.A. Baulin, S. Pogodin, J.Y. Wang, M.J. Tobin, C. Löbbe, R.J. Crawford, Natural bactericidal surfaces: mechanical rupture of Pseudomonas aeruginosa by cicada wings, Small 8 (2012) 2489–2494.
[37] S. Sánchez-Salcedo, M. Colilla, I. Izquierdo-Barba, M. Vallet-Regí, Preventing bacterial adhesion on scaffolds for bone tissue engineering, Int. J. Bioprinting 2 (2015) 20–34.
[38] I. Izquierdo-Barba, M. Colilla, M. Vallet-Regí, Zwitterionic ceramics for biomedical applications, Acta Biomater. 40 (2016) 201–211.
[39] S. Chen, L. Li, C. Zhao, J. Zheng, Surface hydration: principles and applications toward low-fouling/nonfouling biomaterials, Polymer 51 (2010) 5283–5293.
[40] G. Cheng, H. Xue, Z. Zhang, S. Chen, S. Jiang, A switchable biocompatible polymer surface with self-sterilizing and nonfouling capabilities, Angew. Chem. Int. Ed. 47 (2008) 8831–8834.
[41] G. Cheng, G. Li, H. Xue, S. Chena, J.D. Bryers, S. Jiang, Zwitterionic carboxybetaine polymer surfaces and their resistance to long-term biofilm formation, Biomaterials 30 (2009) 5234–5240.
[42] R. Lalani, L. Liu, Electrospun zwitterionic poly(sulfobetaine methacrylate) for nonadherent, superabsorbent, and antimicrobial wound dressing applications, Biomacromolecules 13 (2012) 1853–1863.
[43] J.E. Rosen, F.X. Gu, Surface functionalization of silica nanoparticles with cysteine: a low-fouling zwitterionic surface, Langmuir 27 (2011) 10507–10513.
[44] F. Villegas, L. Garcia-Uriostegui, O. Rodríguez, I. Izquierdo-Barba, A.J. Salinas, G. Toriz, M. Vallet-Regí, E. Delgado, Lysine-grafted MCM-41 silica as an antibacterial biomaterial, Bioengineering 4 (2017) 80.
[45] S. Sánchez-Salcedo, A. García, M. Vallet-Regí, Prevention of bacterial adhesion to zwitterionic biocompatible mesoporous glasses, Acta Biomater. 57 (2017) 472–486.
[46] Z.G. Estephan, J.A. Jaber, J.B. Schlenoff, Zwitterion-stabilized silica nanoparticles: toward nonstick nano, Langmuir 26 (2010) 16884–16889.

[47] Z.G. Estephan, P.S. Schlenoff, J.B. Schlenoff, Zwitteration as an alternative to PEGylation, Langmuir 27 (2011) 6794–6800.
[48] Z.G. Estephan, H.H. Hariri, J.B. Schlenoff, One-pot, exchange-free, room-temperature synthesis of sub-10 nm aqueous, noninteracting, and stable zwitterated iron oxide nanoparticles, Langmuir 29 (2008) 2572–2579.
[49] H. Wei, N.I.J. Lee, H.-S. Han, J.M. Cordero, W. Liu, M.G. Bawendi, Compact zwitterion-coated iron oxide nanoparticles for biological applications, Nano Lett. 12 (2012) 22–25.
[50] H. Wei, O.T. Bruns, O. Chen, M.G. Bawendi, Compact zwitterion-coated iron oxide nanoparticles for in vitro and in vivo imaging, Integr. Biol. 5 (2013) 108–114.
[51] Z. Zhou, L. Wang, X. Chi, J. Bao, L. Yang, W. Zhao, Z. Chen, X. Wang, X. Chen, J. Gao, Engineered iron-oxide-based nanoparticles as enhanced T1 contrast agents for efficient tumor imaging, ACS Nano 7 (2013) 3287–3296.
[52] M. Colilla, I. Izquierdo-Barba, S. Sánchez-Salcedo, J.L.G. Fierro, J.L. Hueso, M. Vallet-Regí, Synthesis and characterization of zwitterionic SBA-15 nanostructured materials, Chem. Mater. 22 (23) (2010) 6459–6466.
[53] I. Izquierdo-Barba, S. Sánchez-Salcedo, M. Colilla, M.J. Feito, C. Ramírez-Santillán, M.T. Portolés, M. Vallet-Regí, Inhibition of bacterial adhesion on biocompatible zwitterionic SBA-15 mesoporous materials, Acta Biomater. 7 (2011) 2977–2985.
[54] M. Colilla, M. Martínez-Carmona, S. Sánchez-Salcedo, M.L. Ruiz-González, J.M. González-Calbet, M. Vallet-Regí, A novel zwitterionic bioceramic with dual antibacterial capability, J. Mater. Chem. B 2 (2014) 5639–5651.
[55] A.G. Gristina, Biomaterial-centered infection: microbial adhesion versus tissue integration, Science 237 (1987) 1588–1595.
[56] J. Gallo, M. Holinka, C.S. Moucha, Antibacterial surface treatment for orthopaedic implants, Int. J. Mol. Sci. 15 (2014) 13849–13880.
[57] H.J. Busscher, H.C. van der Mei, How do bacteria know they are on a surface and regulate their response to an adhering state? PLoS Pathog. 8 (2012) e1002440.
[58] M. Vallet-Regí, M. Colilla, B. González, Medical applications of organic-inorganic hybrid materials within the field of silica-based bioceramics, Chem. Soc. Rev. 40 (2011) 596–607.
[59] S. Lanone, J. Boczkowski, Biomedical applications and potential health risks of nanomaterials: molecular mechanisms, Curr. Mol. Med. 6 (2006) 651–663.
[60] Y. Wang, Q. Zhao, N. Han, L. Bai, J. Li, J. Liu, E. Che, L. Hu, Q. Zhang, T. Jiang, S. Wang, Mesoporous silica nanoparticles in drug delivery and biomedical applications, Nanomed. Nanotechnol. Biol. Med. 11 (2015) 313–327.
[61] M. Vallet-Regí, Bio-Ceramics With Clinical Applications, John Wiley & Sons, Ltd, Chichester, UK, 2014.
[62] S. Wang, Ordered mesoporous materials for drug delivery, Microporous Mesoporous Mater. 117 (2009) 1–9.
[63] M. Manzano, M. Colilla, M. Vallet-Regí, Drug delivery from ordered mesoporous matrices, Expert Opin. Drug Deliv. 6 (2009) 1–18.
[64] P. Yang, S. Gai, J. Lin, Functionalized mesoporous silica materials for controlled drug delivery, Chem. Soc. Rev. 41 (2012) 3679–3698.
[65] M. Vallet-Regí, L. Ruiz-González, I. Izquierdo-Barba, J.M. González-Calbet, Revisiting silica-based ordered mesoporous materials: medical applications, J. Mater. Chem. 16 (2006) 26–31.
[66] M. Vallet-Regí, M. Colilla, I. Izquierdo-Barba, Bioactive mesoporous silicas as controlled delivery systems: application in bone tissue regeneration, J. Biomed. Nanotechnol. 4 (2008) 1–15.

[67] M. Vallet-Regí, J.C. Doadrio, A.L. Doadrio, I. Izquierdo-Barba, J. Pérez-Pariente, Hexagonal ordered mesoporous material as a matrix for the controlled release of amoxicillin, Solid State Ionics 172 (2004) 435–439.
[68] A.L. Doadrio, E.M.B. Sousa, J.C. Doadrio, J. Pérez Pariente, I. Izquierdo-Barba, M. Vallet-Regı´, Mesoporous SBA-15 HPLC evaluation for controlled gentamicin drug delivery, J. Control. Release 97 (2004) 125–132.
[69] J.M. Xue, M. Shi, PLGA/mesoporous silica hybrid structure for controlled drug release, J. Control. Release 98 (2004) 209–217.
[70] I. Izquierdo-Barba, M. Vallet-Regí, N. Kupferschmidt, O. Terasaki, A. Schmidtchen, M. Malmsten, Incorporation of antimicrobial compounds in mesoporous silica film monolith, Biomaterials 30 (2009) 5729–5736.
[71] R. Diez-Martinez, E. Garcia-Fernandez, M. Manzano, A. Martinez, M. Domenech, M. Vallet-Regi, P. Garcia, Auranofin-loaded nanoparticles as a new therapeutic tool to fight streptococcal infections, Sci. Rep. 6 (2016) 19525.
[72] M. Cicuéndez, I. Izquierdo-Barba, M.T. Portolés, M. Vallet-Regí, Biocompatibility and levofloxacin delivery of mesoporous materials, Eur. J. Pharm. Biopharm. 84 (2013) 115–124.
[73] A. Aguilar-Colomer, J.C. Doadrio, C. Perez-Jorge, M. Manzano, M. Vallet-Regi, J. Esteban, Antibacterial effect of antibiotic-loaded SBA-15 on biofilm formation by Staphylococcus aureus and Staphylococcus epidermidis, J. Antibiot. 70 (2017) 259–263.
[74] D. Molina-Manso, M. Manzano, J.C. Doadrio, G. Del Prado, A. Ortiz-Perez, M. Vallet-Regi, E. Gomez-Barrena, J. Esteban, Usefulness of SBA-15 mesoporous ceramics as a delivery system for vancomycin, rifampicin and linezolid: a preliminary report, Int. J. Antimicrob. Agents 40 (2012) 252–256.
[75] L.H. Charlotte, L.L. Wei, S.C. Loo Joachim, Drug-eluting scaffolds for bone and cartilage regeneration, Drug Discov. Today 19 (2014) 714–724.
[76] D. Arcos, A.R. Boccaccini, M. Bohner, A. Díez-Pérez, M. Epple, E. Gómez-Barrena, A. Herrera, J.A. Planell, L. Rodríguez-Mañas, M. Vallet-Regí, The relevance of biomaterials to the prevention and treatment of osteoporosis, Acta Biomater. 10 (2014) 1793–1805.
[77] M. Cicuéndez, J.C. Doadrio, A. Hernández, M.T. Portolés, I. Izquierdo-Barba, M. Vallet-Regí, Multifunctional pH sensitive 3D scaffolds for treatment and prevention of bone infection, Acta Biomater. 65 (2018) 450–461.
[78] R. García-Alvarez, I. Izquierdo-Barba, M. Vallet-Regí, 3D scaffold with effective multidrug sequential release against bacteria biofilm, Acta Biomater. 49 (2017) 113–126.
[79] G.J.A. Boo, D.W. Grijpma, T.F. Moriarty, R.G. Richards, Antimicrobial delivery systems for local infection prophylasis in orthopaedic and trauma surgery, Biomaterials 52 (2015) 113–125.
[80] A.J. Huh, Y.J. Kwon, "Nanoantibiotics": a new paradigm for treating infectious diseases using nanomaterials in the antibiotics resistant era, J. Control. Release 156 (2011) 128–145.
[81] L. Zhang, D. Pornpattananangku, C.M.J. Hu, C.M. Huang, Development of nanoparticles for antimicrobial drug delivery, Curr. Med. Chem. 17 (2010) 585–594.
[82] M. Moritz, M. Geszke-Moritz, The newest achievements in synthesis, immobilization and practical applications of antibacterial nanoparticles, Chem. Eng. J. 228 (2013) 596–613.
[83] X. Li, S.M. Robinson, A. Gupta, K. Saha, Z. Jiang, D.F. Moyano, A. Sahar, M.A. Riley, V.M. Rotello, Functional gold nanoparticles as potent antimicrobial agents against multi-drug-resistant bacteria, ACS Nano 8 (2014) 10682–10686.

[84] K.P. Miller, L. Wang, B.C. Benicewicz, A.W. Decho, Inorganic nanoparticles engineered to attack bacteria, Chem. Soc. Rev. 44 (2015) 7787–7807.
[85] U. Shimanovich, A. Gedanken, Nanotechnology solutions to restore antibiotic activity, J. Mater. Chem. B 4 (2016) 824–833.
[86] S.J. Lam, N.M. O'Brien-Simpson, N. Pantarat, A. Sulistio, E.H.H. Wong, Y.-Y. Chen, J.C. Lenzo, J.A. Holden, A. Blencowe, E.C. Reynolds, G.G. Qiao, Combating multidrug-resistant Gram-negative bacteria with structurally nanoengineered antimicrobial peptide polymers. Nat. Microbiol. 1 (2016) 16162, https://doi.org/10.1038/nmicrobiol.2016.162.
[87] H. Mu, J. Tang, Q. Liu, C. Sun, T. Wang, J. Duan, Potent antibacterial nanoparticles against biofilm and intracellular bacteria. Sci. Rep. 6 (2016) 18877, https://doi.org/10.1038/srep18877.
[88] P.V. Baptista, M.P. McCusker, A. Carvalho, D.A. Ferreira, N.M. Mohan, M. Martins, A.R. Fernandes, Nano-strategies to fight multidrug resistant bacteria—"a battle of the titans" Front. Microbiol. 9 (2018) 1441, https://doi.org/10.3389/fmicb.2018.01441.
[89] Y. Zhao, X. Jiang, Multiple strategies to activate gold nanoparticles as antibiotics, Nanoscale 5 (2013) 8340–8350.
[90] R.R. Castillo, M. Colilla, M. Vallet-Regí, Advances in mesoporous silica-based nanocarriers for co-delivery and combination therapy against cancer, Expert Opin. Drug Deliv. 14 (2017) 229–243.
[91] A. Baeza, M. Colilla, M. Vallet-Regí, Advances in mesoporous silica nanoparticles for targeted stimuli-responsive drug delivery, Expert Opin. Drug Deliv. 12 (2015) 319–337.
[92] M. Martínez-Carmona, M. Colilla, M. Vallet-Regí, Smart mesoporous nanomaterials for antitumor therapy, Nanomaterials 5 (2015) 31906–31937.
[93] F.Q. Tang, L.L. Li, D. Chen, Mesoporous silica nanoparticles: synthesis, biocompatibility and drug delivery, Adv. Mater. 24 (2012) 1504–1534.
[94] C. Fu, T. Liu, L. Li, H. Liu, D. Chen, F. Tang, The absorption, distribution, excretion and toxicity of mesoporous silica nanoparticles in mice following different exposure routes, Biomaterials 34 (2013) 2565–2575.
[95] J. Lu, M. Liong, Z. Li, J.I. Zink, F. Tamanoi, Biocompatibility, biodistribution, and drug-delivery efficiency of mesoporous silica nanoparticles for cancer therapy in animals, Small (16) (2010) 1794–1805.
[96] Y.-S. Lin, N. Abadeer, C.L. Haynes, Stability of small mesoporous silica nanoparticles in biological media, Chem. Commun. 47 (2011) 532–534.
[97] X. Li, Y.-C. Yeh, K. Giri, R. Mout, R.F. Landis, Y.S. Prakash, V.M. Rotello, Control of nanoparticle penetration into biofilms through surface design, Chem. Commun. 51 (2015) 282–285.
[98] D. Pedraza, J. Díez, I. Izquierdo, M. Colilla, M. Vallet-Regí, Amine-functionalized mesoporous silica nanoparticles: a new nanoantibiotic for bone infection treatment, Biomed. Glasses 4 (2018) 1–12.
[99] A. Bandyopadhyay, K.A. McCarthy, M.A. Kelly, J. Gao, Targeting bacteria via iminoboronate chemistry of amine-presenting lipids. Nat. Commun. 6 (2015) 6561, https://doi.org/10.1038/ncomms7561.
[100] A.F. Radovic-Moreno, T.K. Lu, V.A. Puscasu, C.J. Yoon, R. Langer, O.C. Farokhzad, Surface charge-switching polymeric nanoparticles for bacterial cell wall-targeted delivery of antibiotics, ACS Nano 6 (2012) 4279–4287.
[101] S. Brosel-Oliu, N. Abramova, A. Bratov, N. Vigués, J. Mas, F.-X. Muñoz, Sensitivity and response time of polyethyleneimine modified impedimetric transducer for bacteria detection, Electroanalysis 27 (2015) 656–662.

[102] Y. Zhao, Y. Tian, Y. Cui, W. Liu, W. Ma, X. Jiang, Small molecule-capped gold nanoparticles as potent antibacterial agents that target Gram-negative bacteria, J. Am. Chem. Soc. 132 (2010) 12349–12356.

[103] G.R. Newkome, C. Shreiner, Dendrimers derived from 1→3 branching motifs, Chem. Rev. 110 (2010) 6338–6442.

[104] S. Mignani, S.E. Kazzouli, M. Bousmina, J.-P. Majoral, Dendrimer space concept for innovative nanomedicine: a futuristic vision for medicinal chemistry, Prog. Polym. Sci. 38 (2013) 993–1008.

[105] H.W. Van Koten, W.M. Dlakic, R. Engel, M.J. Cloninger, Synthesis and biological activity of highly cationic dendrimer antibiotics, Mol. Pharm. 13 (2016) 3827–3834.

[106] S. Alberti, G.J.A.A. Soler-Illia, O. Azzaroni, Gated supramolecular chemistry in hybrid mesoporous silica nanoarchitectures: controlled delivery and molecular transport in response to chemical, physical and biological stimuli, Chem. Commun. 51 (2015) 6050–6075.

[107] M.A. Mintzer, E.L. Dane, G.A. O'Toole, M.W. Grinstaff, Exploiting dendrimer multivalency to combat emerging and re-emerging infectious diseases, Mol. Pharm. 9 (2012) 342–354.

[108] B. González, E. Ruiz-Hernández, M.J. Feito, C. López de Laorden, D. Arcos, C. - Ramírez-Santillán, C. Matesanz, M.T. Portolés, M. Vallet-Regí, Covalently bonded dendrimer-maghemite nanosystems: nonviral vectors for in vitro gene magnetofection, J. Mater. Chem. 21 (2011) 4598–4604.

[109] B. González, M. Colilla, J. Diez, D. Pedraza, M. Guembe, I. Izquierdo-Barba, M. - Vallet-Regí, Mesoporous silica nanoparticles decorated with polycationic dendrimers for infection treatment, Acta Biomater. 68 (2018) 261–271.

[110] C.Z. Chen, S.L. Cooper, Interactions between dendrimer biocides and bacterial membranes, Biomaterials 23 (2002) 3359–3368.

CHAPTER FOUR

Tumor Targeting and Tumor Growth Inhibition Capability of Mesoporous Silica Nanoparticles in Mouse Models

Tan Le Hoang Doan*,†,‡, Ngoc Xuan Dat Mai*,†, Kotaro Matsumoto*, Fuyuhiko Tamanoi*,§,1

*Institute for Integrated Cell-Material Sciences, Institute for Advanced Study, Kyoto University, Kyoto, Japan
†Center for Innovative Materials and Architectures (INOMAR), University of Science, Vietnam National University—Ho Chi Minh City, Ho Chi Minh City, Vietnam
‡Faculty of Chemistry, University of Science, Vietnam National University—Ho Chi Minh City, Ho Chi Minh City, Vietnam
§Department of Microbiology, Immunology and Molecular Genetics, University of California, Los Angeles, CA, United States
[1]Corresponding author: e-mail address: fuyut@microbio.ucla.edu

Contents

Abstracts

Mesoporous silica nanoparticles (MSNs) provide a promising vehicle for anticancer drug delivery. Various animal studies point to the efficacy of this nanoparticle for delivering anticancer agents (drug and siRNA) to inhibit tumor growth. These studies also showed tumor accumulation of MSN nanoparticles. While the extent of tumor accumulation differed, the study showed that it is possible to achieve significant accumulation of nanoparticles in the tumor. Biocompatibility and safety of MSN were also demonstrated by these studies.

The Enzymes, Volume 44
ISSN 1874-6047
https://doi.org/10.1016/bs.enz.2018.09.002
© 2018 Elsevier Inc.
All rights reserved.

1. INTRODUCTION

Mesoporous silica nanoparticles (MSNs) have emerged as a promising type of nanomaterial for cancer therapy [1]. Because of the presence of large surface area, various anticancer drugs can be loaded onto the nanoparticles. MSNs are relatively stable enabling a variety of chemical modifications to take place [2].

There are a significant number of reports that demonstrate excellent tumor growth inhibition when MSNs were used to deliver anticancer agents such as drugs and siRNA. MSNs exhibit highly efficient loading capacity with anticancer drugs such as doxorubicin and gemcitabine owning to their large pore size structure. Moreover, the drug release of MSNs can be controlled by using nanovalve structure which is able to response by internal or external conditions. Camptothecin delivery by using phosphonate modified MSN was effective in inhibiting pancreatic cancer xenografts [3,4]. Use of lipid bilayer coated MSN was also quite effective in inhibiting pancreatic cancer in the K-ras pancreatic cancer model [5]. In addition, hollow MSN was used to inhibit tumor growth in mouse studies [6–8]. The delivery of TWIST siRNA shuts down expression of the TWIST gene, a master regulator of the EMT transition, and results in the inhibition of tumor growth. Moreover, this sensitizes ovarian cancer tumor to cisplatin reflecting the role of TWIST inhibition to reverse drug resistance [9–11].

Various animal studies described above are accompanied by the inquiry of tumor accumulation of MSN nanoparticles. Examination of these results leads to a conclusion that it is possible to accomplish tumor accumulation of MSNs. While the extent of tumor accumulation differs among different experiments, clear signal of MSN localization in the tumor can be obtained. The key is to increase this tumor accumulation. Distribution of MSNs in other organs is also detected and we note that liver, spleen, and kidney localization is widely observed. Further work is needed to investigate what determines these distribution profiles. Size and surface charge are two examples of factors that could influence biodistribution. In addition, the amount of MSN injected has a profound effect on biodistribution.

Animal studies also provided results in support of the idea that MSNs are biocompatible and safe. Various studies report reduced toxic effect of anticancer drugs when delivered by MSNs. Short term as well as long term studies were carried out examining liver and kidney profile as well as blood profile. Recently, a novel type of MSNs can be designed for biodegradation to avoid long-term accumulation. These particles contain biodegradable bonds such as peptide, disulfide, and tetrasulfide within the framework structure leading to capability of their decomposition for few days in body fluid media [12–15]. However, the new particles have not been studied in mouse models.

In this chapter, we review some representative mouse studies that address the above points.

2. MSNs WITH PHOSPHONATE SURFACE MODIFICATION

Phosphonate surface modified MSNs were shown to be effective in delivering anticancer drugs and inhibit tumor growth in mouse xenograft studies [3,4]. This surface modification generating negatively charged MSNs was initially carried out to increase nanoparticle dispersion and to prevent aggregation. The size of the nanoparticles was about 130 nm and the particle contained 1400 pores and each pore had a diameter of 2–3 nm (Fig. 1A).

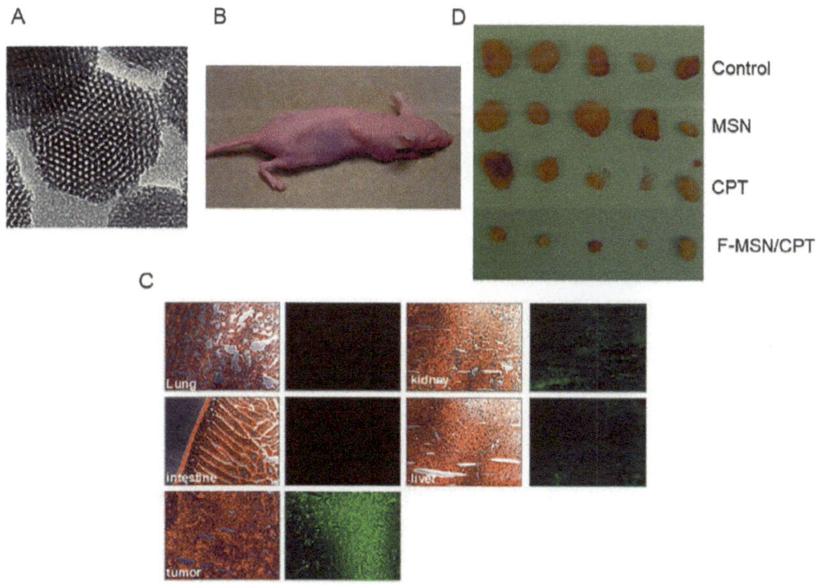

Fig. 1 Tumor growth inhibition capability of phosphonate modified MSN. (A) Transmission electron microscopy images of MSNs; (B) nude mice with xenografts of human breast cancer cell MCF-7; (C) biodistribution of FITC labeled FMSNs in mice with xenograft tumors; (D) representative photos of tumor tissues collected after the treatment with camptothecin loaded MSN (F-MSN/CPT), MSN or CPT. Control is no treatment. *Adapted from J. Lu, M. Liong, Z. Li, J.I. Zink, F. Tamanoi, Biocompatibility, biodistribution, and drug-delivery efficiency of mesoporous silica nanoparticles for cancer therapy in animals, Small 6 (16) (2010) 1794–1805.*

Maximum tolerated dose of MSNs was determined by injecting increasing amount of the phosphonate-modified MSNs into wild type mice at an amount that ranged from 0.06 to 1 mg per mouse. The results showed that even at 1 mg per mouse (50 mg/kg), the MSNs were well tolerated, as little adverse effect was observed with blood profile, kidney and liver enzyme profiles. We did note minor changes with some mice at different time points including mild increase in neutrophil and eosinophil count as well as slightly elevated liver enzyme. Based on these results, the amount of MSN injected into a mouse was kept below 1 mg per mouse in subsequent experiments. Excretion of MSN into feces and urine was observed. Within 4 days after injection, almost all MSNs were recovered outside the animal body.

Tumor accumulation of the phosphonate modified MSNs was demonstrated by injecting MSNs intravenously into mice with breast cancer xenograft (Fig. 1B and C). Initial experiment was carried out using whole body imaging with MAESTRO after intravenous injection of fluorescent MSNs. This revealed excellent accumulation of MSNs in the tumor. Analysis of fluorescence in thin section from the tumor as well as from various organs showed significant fluorescence in the tumor. In addition to the tumor, significant level of fluorescence was observed in kidney and liver. On the other hand, little fluorescence was observed with intestine. Biodistribution was further investigated using ICP-MS of Si that provided more quantitative analysis.

Finally, efficacy of phosphonate surface modified MSNs to inhibit tumor growth was demonstrated using camptothecin (CPT) loaded MSN and by employing MCF-7 mouse xenograft model. Significant tumor growth inhibition was observed after twice per week injection for 50 days. Tumor growth inhibition efficacy of CPT-loaded MSN was further investigated using two different human pancreatic cancer xenografts, PANC-1 and MiaPaCa-2 xenografts. With the PANC-1 xenograft model, significant tumor growth inhibition was observed after twice per week injection for 50 days (Fig. 1D) confirming the earlier result with MCF-7 xenograft. Better results were obtained by using MiaPaCa-2 xenografts; tumor regression was observed after injecting CPT-loaded MSN for 1 week (twice per week) and the tumor was virtually eliminated. In all these experiments, no significant adverse effect was detected during the course of the experiment.

Contribution of folate modification of MSN was examined. In both the above pancreatic cancer xenograft models, we observed significant improvement of tumor growth inhibition capability with folate modified MSN.

On the other hand, little difference was observed between folate modified and nonmodified MSNs using the MCF-7 xenograft. The effect of folate targeting may differ among different cancer models.

3. MSNs WITH LIPID BILAYER COATING (PROTOCELL)

3.1 Protocell

MSNs with lipid bilayer coating was developed with the idea to combine attractive features of liposomes (high biocompatibility, flexible formulation, low immunogenicity) with stability of MSN [16,17]. To create a protocell, MSN cores were covered by supported lipid bilayer and various ligands are attached. Liu et al. [17] described a system where liposome fusion on a mesoporous silica particle core simultaneously loads and seals the cargo. This involves fusion of a positively charged lipid bilayer such as DOPC/DOTAP or DOPC/DOPS on a negatively charged mesoporous silica core. The average particle size was ~100 nm. In this system, loading of the negatively charged drug is controlled by the lipid composition. For example, loading of calcein increased in correspondence with the percentage of positively charged DOTAP in contrast with negatively charged DOPS. The protocells can be taken up into Chinese hamster ovary (CHO) cells and calcein was released inside the cell.

3.2 Silicasome Platform

The protocell approach was further adapted to provide an effective platform for high-dose pancreatic ductal adenocarcinoma (PDAC) chemotherapy. This platform that uses MSNs covered with lipid bilayer (LB-MSN) is termed "silicasome" (Fig. 2A shows one example) [5].

Efficacy of the silicasome platform to deliver irinotecan and to inhibit tumor growth was examined using a pancreatic cancer mouse model [19]. In the experiment shown in Fig. 2C, irinotecan delivery by the silicasome platform was effective in inhibiting growth of pancreatic cancer in mouse models [19]. The model used is KPC-derived orthotopic model that mimics human PDAC in the K-ras mutation. Intravenous injection of irinotecan loaded LB-MSN every 4 days for up to 8 weeks resulted in a significant inhibition of tumor growth. The efficacy was better than the free drug or a liposomal drug. Histological evaluation of liver from treated animals showed that the toxicity due to irinotecan was significantly reduced in the LB-MSN formulation. Biodistribution of LB-MSN was examined

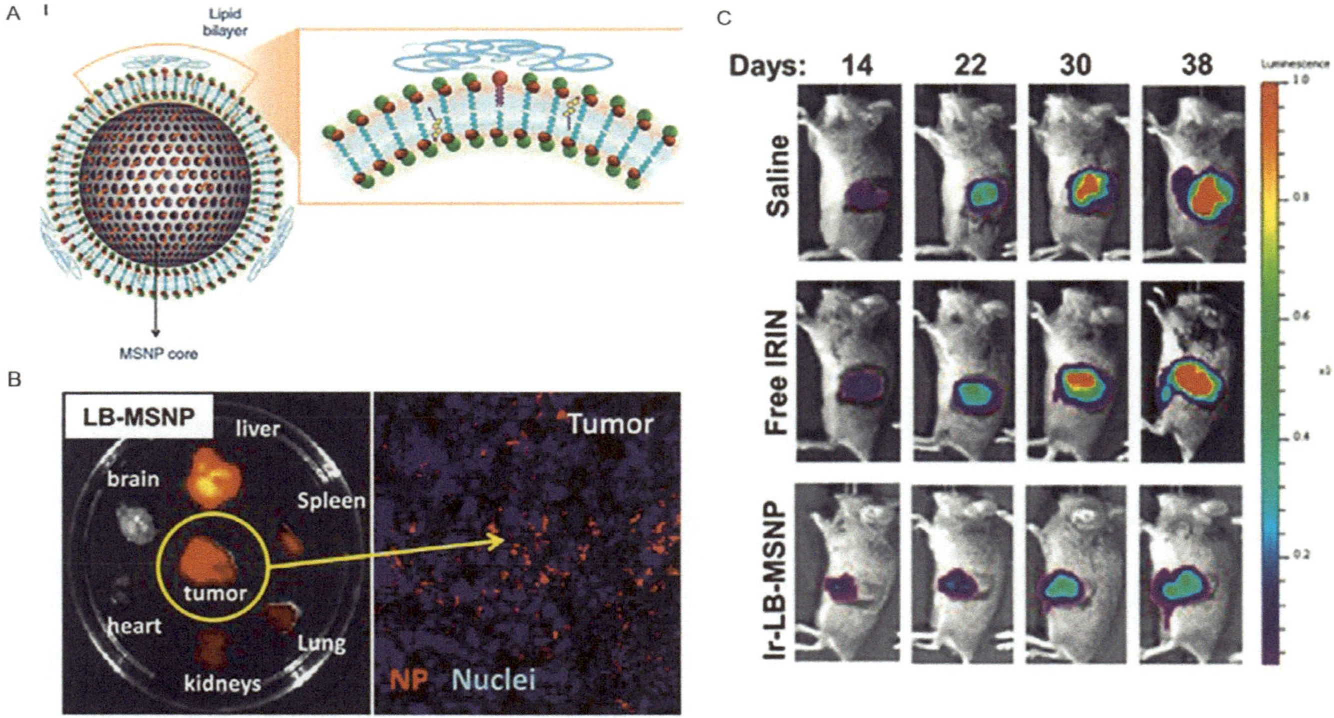

Fig. 2 Tumor growth inhibition capability of lipid bilayer coated MSN (silicasome). (A) A schematic to show the structure of an example of lipid bilayer MSNP (LB-MSNP). See Ref. [18] for more detail; (B) biodistribution of NIR-labeled LB-MSN. Tumor tissues as well as major organs (heart, lung, spleen, liver, kidney, and brain) were collected for ex vivo imaging after Ir-LB-MSNP administration; (C) tumor growth inhibition representative IVIS imaging of luciferase expressing tumor on days 14, 22, 30, and 38, tumor burden was plotted as fold-increase compared to the nontreated control. *Adapted from X. Liu, A. Situ, Y. Kang, K.R. Villabroza, Y. Liao, C.H. Chang, T. Donahue, A.E. Nel, H. Meng, Irinotecan delivery by lipid-coated mesoporous silica nanoparticles shows improved efficacy and safety over liposomes for pancreatic cancer, ACS Nano 10 (2) (2016) 2702–2715; J. Lu, X. Liu, Y.-P. Liao, F. Salazar, B. Sun, W. Jiang, C.H. Chang, J. Jiang, X. Wang, A.M. Wu, H. Meng, A.E. Nel, Nano-enabled pancreas cancer immunotherapy using immunogenic cell death and reversing immunosuppression, Nat. Commun. 8 (1) (2017) 1811.*

using NIR (near infrared labeled Dylight 680) labeled nanoparticle. Significant accumulation in the tumor was observed (Fig. 2B).

The above approach was further improved by coadministration of free iRGD peptide that enhanced carrier uptake (three- to fourfolds) [5]. This cyclic peptide can trigger NRP-1 (neurotrophin-1) mediated transcytosis. Coadministration of iRGD peptide with LB-MSN loaded with irinotecan resulted in shrinking the primary tumor and inhibited metastasis to liver, stomach, and intestine in the K-ras orthotopic model. This resulted in improved survival.

Lu et al. [18] used the silicasome platform to develop a strategy to induce immunogenic cell death and reverse immunosuppression. This was accomplished by loading a prodrug of indoximod (IND-PL) into lipid bilayer and oxaliplatin (OX) into the porous interior of LB-MSN. OX is a chemotherapy drug that induces immunogenic cell death (ICD). IND is an inhibitor of the immunosuppressive indoleamine 2,3-dioxygenase (IDO) pathway. The particles, size of ~83 nm, are uniformly coated with an entire LB, ~6.5 nm thick, and have a slightly negative zeta potential. Its stability in biological media is up to 30 days with about 4% of OX loading. Administration of the silicasome loaded with IND-PL and OX into the KPC mouse model resulted in enhanced tumor shrinkage and inhibition of metastatic spread. There was significant survival benefit by the treatment. Biodistribution of OX/IND MSN labeled with NIR dye revealed accumulation in the tumor that was retained for up to 48 h.

4. MSNs COATED WITH PEI USED FOR DRUG AND siRNA DELIVERY

PEI-coated MSN has been developed to delivery nucleic acids such as siRNA. Hom et al. [20] showed that siRNA can be delivered to cancer cells. Xia et al. [21] characterized the effect of PEI surface modification to increase uptake into cancer cells and developed MSN that has favorable property as a delivery vehicle. This involves the use of copolymer including PEI and polyethylene glycol (PEG) [22–24]. The zeta-potential measurement showed the positive-charged surface of the copolymer coating MSN about +45 mV. The effect of the particle size with range of 50–100 nm on in vivo delivery was studied on murine xenograft model in nude mice. As a result of in vivo doxorubicin delivery, the copolymer coated MSN with particle size of 50 nm termed NP3 showed the best tumor accumulation and tumor growth inhibition (Fig. 3A and B) [22]. The particle was distributed also to the organs, especially

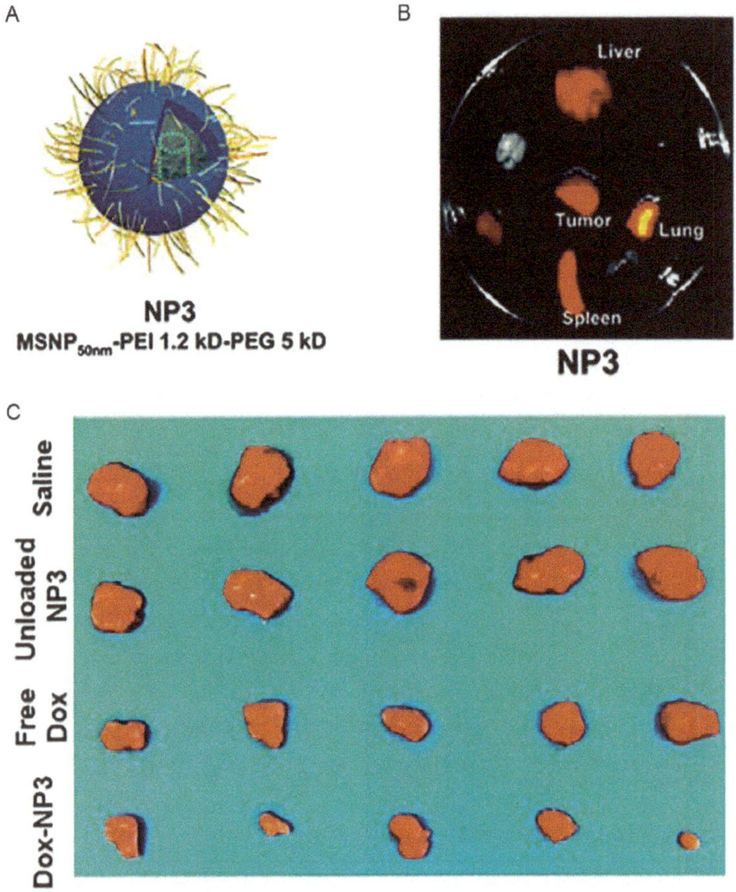

Fig. 3 Tumor growth inhibition of PEI-coated MSN against pancreatic cancer. (A) Schematic illustration of the structure of PEI-coated MSN (NP3); (B) biodistribution of NIR-labeled NP3 in tumor-bearing nude mice after injection (72 h). Tumor tissues as well as major organs (heart, lung, spleen, liver, kidney, brain, and muscle) were collected for ex vivo imaging in fluorescence microscope; (C) tumor growth inhibition by the treatment with doxorubicin loaded NP3. Tumor tissues collected after the treatment with Dox-NP3, unloaded NP3, free Dox or saline. *Adapted from H. Meng, M. Xue, T. Xia, Z. Ji, D.Y. Tarn, J.I. Zink, A.E. Nel, Use of size and a copolymer design feature to improve the biodistribution and the enhanced permeability and retention effect of doxorubicin-loaded mesoporous silica nanoparticles in a murine xenograft tumor model, ACS Nano 5 (5) (2011) 4131–4144.*

liver, spleen, and lung. Doxorubicin loaded NP3 was effective in inhibiting tumor growth in a pancreatic cancer xenograft (Fig. 3C).

The use of PEI-coated MSN enables dual delivery of anticancer drug and siRNA, as drug molecules can be stored in the pores while siRNA can be bound to PEI surrounding MSN. Meng et al. [23] used siRNA against MDR1, that is responsible for multidrug resistance mechanism. Dox and siRNA loaded MSN was intravenously injected every 3–6 days for 30 days into nude mice transplanted with MCF-7/MDR (a doxorubicin resistant clone of MCF-7) cells. This resulted in significant inhibition of tumor growth. Side effect of administering free doxorubicin was not observed with the MSN treatment.

PEI-PEG coated MSN was employed to develop a therapy against pancreatic cancer [24]. In this approach, the PEI coating enabled molecular complexation to a small molecule inhibitor of TGF-β that can lead to optimal passive delivery of liposomes containing gemcitabine, a first-line chemotherapeutic agent. The nitrogen atom of the TGF-β inhibitor LY364947 can form a hydrogen bond with noncomplexed hydrogen of PEI. The TGFβi-MSN improved intratumoral distribution of liposome in the BxPC3 xenograft model. Xenograft bearing nude mice were intravenously injected first with TGFβi-MSN followed 2h later by the injection of liposomes containing gemcitabine. This paired set of injection was repeated every 3–6 days for 38 days resulting in a significant inhibition of tumor growth. This type of treatment is termed "two-wave therapy."

5. PEI-COATED MSNs FOR TWIST siRNA DELIVERY

Finlay et al. [9] have shown that MSN coated with low molecular weight polyethyleneimine (PEI) provides an effective vehicle for interfering RNA (siRNA) therapeutics, which is capable of posttranscriptional gene silencing in a sequence specific process [9–11]. In their approach, the *TWIST* gene was chosen as a target of gene shutdown. This gene is a master regulator of the EMT (epithelial-mesenchymal transition) that controls angiogenesis, metastasis, stemness and drug resistance. The PEI-coated MSN had a size of 127 nm based on TEM measurement and positive surface charge with the zeta potential +43.75 mV. The siRNA against the *TWIST1* gene was synthesized with various chemical modifications to increase efficacy and stability. In xenograft mouse model investigation, the siRNA delivery via PEI-coated MSN by intraperitoneal (IP) could knock down the expression of the transcriptions leading to reduced tumor burden. The *TWIST* gene was knocked down in the tumor. In addition, other EMT genes under control of TWIST such as Vimentin were also knocked down in the tumor.

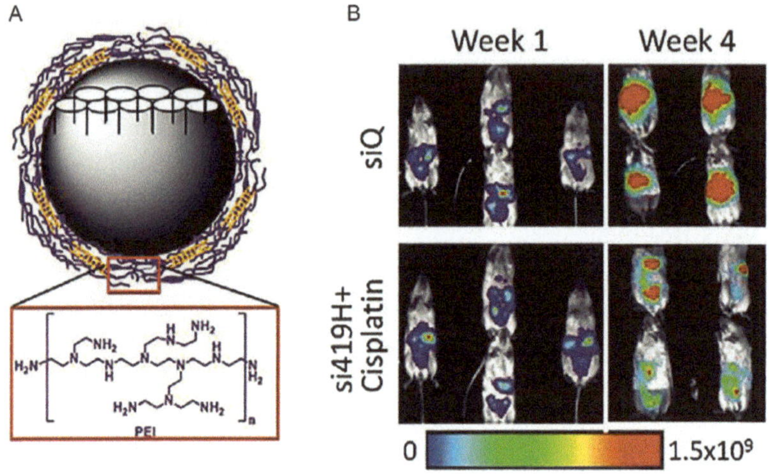

Fig. 4 TWIST siRNA delivery using PEI-coated MSN against ovarian cancer. (A) MSNs used in these studies have a PEI (structure is shown below) coating (purple layer) which binds siRNA (orange); (B) tumor growth inhibition by the administration of siTWIST loaded MSN was monitored by the bioluminescence of ovarian tumor formed on NSG mice. Note that the tumor enlargement is seen with the control (control siQ RNA loaded MSN) as detected by the increased bioluminescence signal, while the tumor enlargement is much less with the combined treatment with siTWIST loaded MSN and cisplatin. *Adapted from C.M. Roberts, S.A. Shahin, W. Wen, J.B. Finlay, J. Dong, R. Wang, T.H. Dellinger, J.I. Zink, F. Tamanoi, C.A. Glackin, Nanoparticle delivery of siRNA against TWIST to reduce drug resistance and tumor growth in ovarian cancer models, Nanomedicine 13 (3) (2017) 965–976.*

This initial study was followed by the demonstration that shutting down *TWIST* sensitizes tumor to cisplatin in the ovarian cancer mouse model [10]. This animal model was established by intraperitoneal injection of OVCAR8-IP cells into NSG mice (Fig. 4). One week after the inoculation of OVCAR8-IP cells, once or twice per week IP injection of siRNA loaded MSN together with cisplatin was carried out for a total of 4 weeks. The combined use of TWIST siRNA-MSN and cisplatin led to a dramatic inhibition of tumor growth (Fig. 4B). More recently, significant improvement was observed by introducing hyaluronic acid (HA) targeting to MSN (MSN-HA) [11]. MSN-HA was prepared by first synthesizing PEI-coated MSN and then covalently bonding HA (hyaluronic acid) to the amine groups in the PEI using EDC-NHS coupling reaction. HA binds CD44 and could target CSCs (cancer stem cells).

Epithelial ovarian cancer (EOC) is one of the leading causes of cancer-related deaths among women worldwide. There is a need for effective therapy against EOC. Thus, these studies provide a promising avenue to develop a novel therapy against EOC. These studies are described in the chapter "Nanoparticle delivery of TWIST small interfering RNA and anticancer drugs: A therapeutic approach for combating cancer" by Glackin. Please refer to that chapter for more discussion.

Excellent tumor accumulation of siRNA was noted in the above experiment using PEI-MSN [10]. In this work, intraperitoneal (IP) administration of OVCAR8-IP cells into mice successfully produced primary tumors (localized to the ovary) and disseminated masses within the peritoneal cavity. IP injection of MSN carrying fluorescent (Alexa Fluor 647) siRNA showed that the MSN carrying siRNA selectively accumulated in the primary tumors as well as in disseminated mases. Accumulation in other peritoneal tissue or organ was much less.

6. CARBOXYL FUNCTIONALIZATION

A different type of negatively charged MSNs was examined by Xie et al. [25]. This type of MSNs was synthesized so that they are functionalized with carboxyl group on the pore surface. The average diameter was 55 nm and a zeta potential was −20 mV. Cy5 dye was attached to investigate biodistribution of the nanoparticle. Injection of these nanoparticles via tail vein into nude mice revealed quick accumulation in the liver and this was maintained for 2 days. In addition, distribution in the spleen and kidney could be seen.

Antitumor effect of these nanoparticles loaded with doxorubicin demonstrated significant effect on hepatocellular carcinoma. Doxorubicin loaded MSN/COOH was intravenously injected into the tumor model of ICR mice bearing H22 tumor. Significant enhancement of the antitumor effect of doxorubicin was observed with the MSN/COOH loaded with doxorubicin. The authors suggested use of these nanoparticles for liver cancer therapy.

7. HOLLOW MSN

Hollow mesoporous silica nanoparticle is another type that has been studied recently [6–8,26,27]. As with other MSNs, HMSN can be functioned and surface modified with some specific factors to achieve effective targeting.

Chen et al. [6] fabricated a well-functionalized HMSN nanoconjugates by multistep surface modification of HMSN, including amino group surface functionalization, anticancer drug (doxorubicin) loading, near infrared (NIR) dye such as ZW800 and ^{64}Cu-chelator for PET imaging. Afterward, the nanoparticles were PEGylated and anti-CD105 antibody [28] was attached. This ^{64}Cu-HMSN-ZW800-TRC105 has uniform size of ~150 nm. Due to hollow space inside HMSN, significant enhancement of drug loading was achieved. In vivo tumor targeting efficacy was evaluated using 4T1 murine breast tumor-bearing mice by injecting intravenously the nano-conjugated via tail vein (10 mg/kg). Its accumulation in the T41 tumor was noted. In addition to tumor accumulation, nanoparticle also distributed in liver and spleen.

Li et al. [29] synthesized a type of mesoporous and hollow silica nanoparticles, called silica nanorattles with various size from 60 to 580 nm. After PEGylation, all the nanoparticles had near-neutral zeta potentials, and their morphology and diameter were unchanged. The in vivo distribution of particles was investigated by injecting into subcutaneous hepatic H22 tumor-bearing mice via the tail vein (a dose of 40 mg/kg). Based on fluorescence examination, all the PEG-MHSNs had a distribution in liver and the signal intensity was dependent on particle size. A similar result also obtained in the spleen, increasing distribution with increase of particle size. In contrast, no fluorescent signal was recorded for PEG-HMSN particle in kidney, except a small amount of 120 nm particles. Tumor accumulation was highest with 120 nm PEG-MHSN-TRITC.

To gain better tumor targeting efficacy, poly(ethylene glycol)-folic acid (PEG-FA) was introduced into hollow structure of MSNs [30]. Via a pH-sensitive benzoic-imine covalent bond, DOX itself was used to cap the outlets of channels. Furthermore, siRNA against P-gp was codelivered to downregulate P-gp expression and inhibit drug resistant human breast cancer cell. The spherical MSNs-siRNA@DOX-PEG-FA had a diameter around 160 nm with surface charge of -8.65 ± 0.57 mV. The tumor model was generated on the dorsal of the mice by injecting MCF/ADR cells. Free DOX or DOX-loaded NPs were intravenously injected via tail vein and in vivo biodistribution, after 3 h of injection was examined (Fig. 5A), which revealed that DOX level was high in tumor, but was also detected in liver and kidney. The DOX fluorescence intensity in liver and kidney reduced at 24 h postinjection. The DOX tumor accumulation with both MSNs-siRNA-@DOX-PEG and MSNs-siRNA@DOX-PEG-FA was higher than free DOX. Importantly, no obvious changes in the morphology of major organs were observed for all DOX-loaded formulations, which demonstrated the excellent antitumor ability and low toxicity of this codelivery. Administration of the hollow MSN padded with siRNA and DOX resulted in significant inhibition of xenograft tumor growth (Fig. 5B).

A tumor-targeting drug delivery nanoplatform of HMSN was recently developed by Zhou et al. [7] (Fig. 6). Transferrin was conjugated onto the

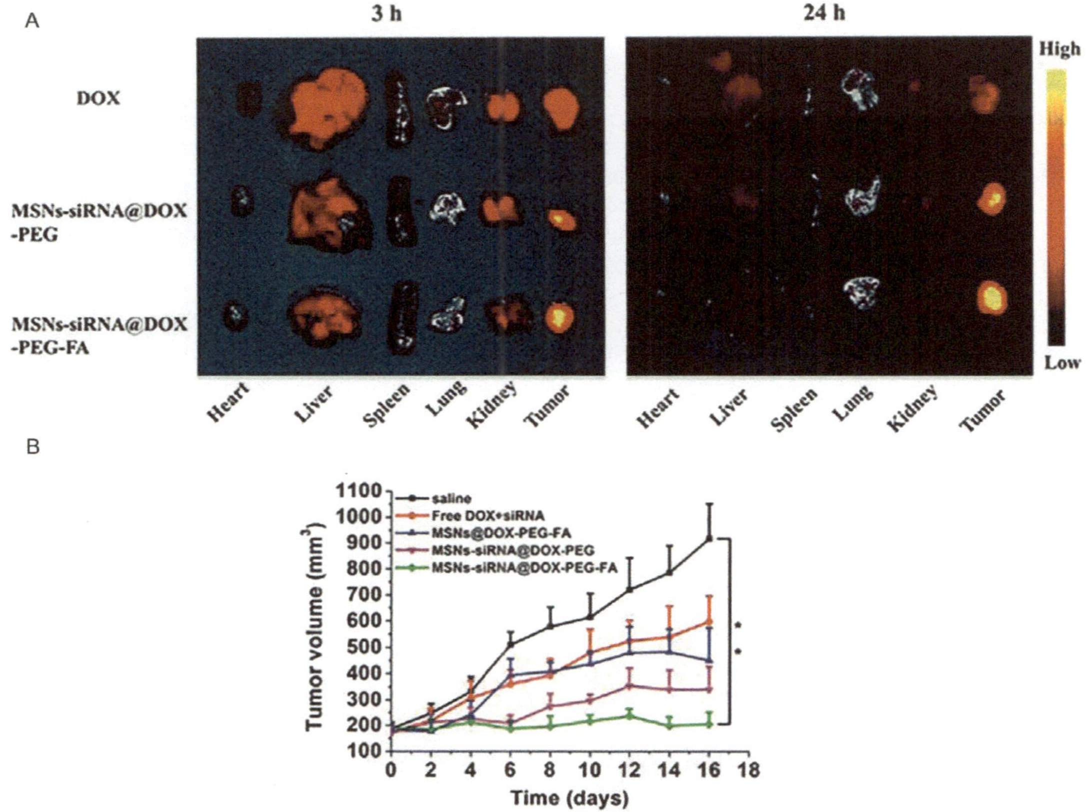

Fig. 5 Tumor growth inhibition capability of PEG-FA modified hollow MSN. (A) Doxorubicin fluorescence in tumor and various organs after intravenous injection of DOX loaded PEG-FA HMSN, DOX loaded PEG MSN or free DOX. Tumor tissues and major organs (heart, lung, spleen, liver, kidney, and brain) were collected for ex vivo fluorescence imaging; (B) in vivo antitumor efficacy of DOX loaded PEG-FA HMSN as well as other constructs by systemic injection into tumor SCID mice. *Adapted from W. Cheng, C. Liang, X. Wang, H.-i. Tsai, G. Liu, Y. Peng, J. Nie, L. Huang, L. Mei, X. Zeng, A drug-self-gated and tumor microenvironment-responsive mesoporous silica vehicle: "four-in-one" versatile nanomedicine for targeted multidrug-resistant cancer therapy, Nanoscale 9 (43) (2017) 17063–17073.*

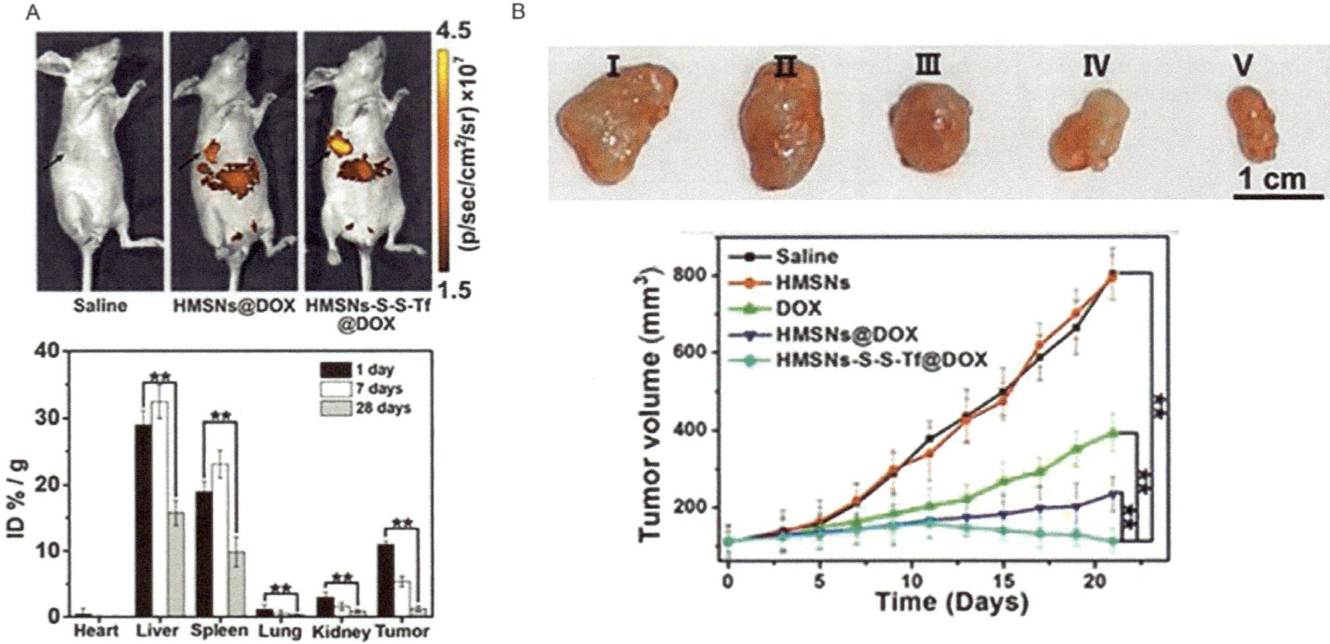

Fig. 6 Tumor growth inhibition capability of transferrin conjugated hollow MSN. (A) DOX-conjugated HMSN was injected into nude mice bearing MDA-MB-231 tumor and the appearance of DOX fluorescence in the tumor was examined as a measure of biodistribution. Note that the fluorescence in the tumor is higher with HMSN with transferrin than the construct without transferrin. The lower panel shows biodistribution at different time intervals. (B) Inhibition of tumor growth by administering DOX-conjugated HMSN for 21 days. Various treatments used are described in the figure. *Adapted from J. Zhou, M. Li, W.Q. Lim, Z. Luo, S.Z.F. Phua, R. Huo, L. Li, K. Li, L. Dai, J. Liu, K. Cai, Y. Zhao, A transferrin-conjugated hollow nanoplatform for redox-controlled and targeted chemotherapy of tumor with reduced inflammatory reactions, Theranostics 8 (1) (2018) 518–532.*

surface via disulfide linkage resulting in the removal from the particle in the intracellular redox condition in the tumor. As a consequence, DOX was released from the nanocarrier causing inhibition of tumor growth. Moreover, transferrin corona could increase blood circulation of nanoplatform as well as diminish the inflammatory response after intravenous administration in vivo. The average diameter of HMSN-S-S-Tf@DOX slightly rose to 167 nm after conjugation compared to original HMSNs and still maintained the monodispersity. The MDA-MB-231 tumor growth in nude mouse models was used to investigate the therapeutic effect. HMSNs@DOX and HMSNs-S-S-Tf@DOX nanoplatforms displayed excellent tumor growth inhibitory efficiency. Excellent tumor-targeting capability in vivo of HMSNs-S-S-Tf was revealed by greater intensity of DOX fluorescence in the tumor compared with that of HMSN@DOX (Fig. 6A). Major organs from the tumor-bearing mice were also collected and stained with H&E for histological analysis. While the administration of free DOX resulted in severe myocardial damage, no significant histological changes in the sections of organs from all of nanocarriers. Tumor from mice after treatment with the HMSN for 21 days was much smaller than that of mice treated with empty HMSN (Fig. 6C). These results suggest that this nanoplatform has a promising therapeutic capability.

8. MSNs WITH NANOVALVES

Controllable release of anticancer drugs is an important property to enhance efficiency and reduce toxic side effect of nanodelivery system [1]. This feature allows nanoparticles to keep the drug inside their structure and release the cargo at target tumor only when they receive a specific signal to release [31]. Among many nanomaterials designed for drug-controlled release capability, nanovalves-based MSN has emerged as a promising carrier for anticancer drug delivery application because of their advantages including prevention of any premature release. The main strategy for nanovalve MSN synthesis is a "capping" and "gating" approach in which organic molecules such as rotaxanes and pseudorotaxanes are attached to the pore openings to provide open/close function for the cargo in the pore [32]. The drug release from the nanovalves could be stimulated by internal conditions such as pH [33–39], redox state [40–45], enzymes [46–48], or biomolecules [49–51] as well as by external activation such as light [52–57] or magnetic field [58–63].

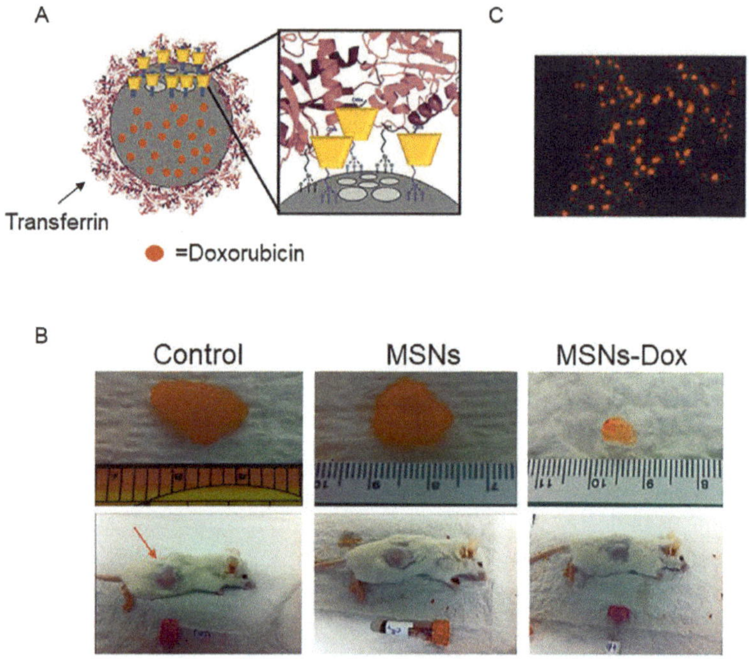

Fig. 7 Tumor growth inhibition of nanovalve equipped MSN. (A) Schematic overview of nanovalve-equipped MSN. The nanoparticle is equipped with pH sensitive nanovalve and is also surface modified by transferrin; (B) inhibition of tumor growth after administering DOX loaded MSN or empty MSN. Control was not treated with nanoparticles; (C) doxorubicin fluorescence in the tumor after the treatment demonstrating successful delivery of DOX. *Adapted from A.A. Hwang, J. Lu, F. Tamanoi, J.I. Zink, Functional nanovalves on protein-coated nanoparticles for in vitro and in vivo controlled drug delivery, Small 11 (3) (2015) 319–328.*

The first nanovalve MSN employed for in vivo drug delivery was reported by Lu et al. [64] (Fig. 7). The pH-sensitive nanovalve MSN system was composed of an aniline-based stalk grafted onto the surface of MSN, which was capped with α-cyclodextrin to seal cargo inside the pores at neutral pH conditions (Fig. 7A). The nanovalve could open and deliver the cargo when exposed to a pH lower than 6. Moreover, the nanovalve MSN surface was covalently attached to transferrin as an active targeting agent via mercapto-modification to generate nanoparticles named Tf-NV-MSN. The in vivo drug delivery experiments in xenograft MiaPaCa-2 (pancreatic cancer cell) tumor demonstrated that the doxorubicin loaded Tf-NV-MSN was efficient in inhibiting tumor growth and releasing the drug in the tumor (Fig. 7B and C).

Dai et al. [65] developed a redox-responsive nanovalve material based on yolk-shell MSN (YMSN) whose structure contained interior hollow spaces and permeable outer shell. The nanovalve YMSN was synthesized via a surfactant assembly sol–gel reaction between TEOS and 1,2-bis(triethoxysilyl) ethane, followed by fabrication of Pd templated rotaxane to from a nanovalve through redox-breakable disulfide linkers. Folic acid was conjugated on the nanovalve MSN as targeting motifs. The average particle size was about 100 nm and the surface charge values were −1.4 mV. For study of in vivo drug delivery, a HEPG2 (liver cancer cell) tumor-bearing nude mouse model was employed to evaluate the antitumor effect of doxorubicin loaded YMSN via intravenous injection. The highly effective tumor inhibition of YMSN@Dox was demonstrated by monitoring tumor volumes and weights. The results showed that the DOX loaded nanoparticle inhibited the tumor, while there was no obvious damage to major organs (lung, heart, spleen, liver, and kidney).

The tumor targeted drug delivery in vivo of an enzyme responsive nanovalve MSN was reported by Gayam et al. [66]. The nanovalve system was constructed by interlocked molecules in the form of rotaxane in which an α-cyclodextrin was bound onto a diethylene glycol stalk. The stalk was attached on an MSN surface with benzoquinone stopper. The NAD(P)H:quinone oxidoreductase 1 (NQO1) enzyme was served as a stimulus for drug release of the system (named MSN-NQO1). The average size of MSN-NQO1 was estimated in range of 90–110 nm. Zeta potential measurement showed the positive surface charge of the nanoparticle about +27.51 mV. The antitumor efficiency of doxorubicin loaded MSN-NQO1 was studied on A549 (lung cancer cell) xenograft tumor in vivo. The result in the tumor volume after therapy for 14 days in nude mice showed the higher tumor inhibition with the Dox loaded nanovalve system than the free Dox and saline systems.

9. FUTURE PROSPECTS

A number of mouse studies suggest that MSNs can deliver anticancer agents and inhibit tumor growth. MSN itself does not have effect suggesting nontoxic nature of the vehicle when used in amounts below maximum tolerated dose (MTD). It is, however, necessary to define MTD when a new material is to be used. The amount of MSN to achieve tumor accumulation also needs to be calibrated. If an optimum amount is defined, significant tumor accumulation can be achieved.

ACKNOWLEDGMENTS

This work was supported by grants from JSPS KAKENHI grant number JP15K21764 and from Vietnam National University Ho Chi Minh City grant number VNU-A-562-2018-50-01. We thank Hiroko Imai for her assistance in the preparation of the manuscript.

REFERENCES

[1] H. Mekaru, J. Lu, F. Tamanoi, Development of mesoporous silica-based nanoparticles with controlled release capability for cancer therapy, Adv. Drug Deliv. Rev. 95 (2015) 40–49.

[2] I.I. Slowing, B.G. Trewyn, S. Giri, V.S.-Y. Lin, Mesoporous silica nanoparticles for drug delivery and biosensing applications, Adv. Funct. Mater. 17 (8) (2007) 1225–1236.

[3] J. Lu, M. Liong, Z. Li, J.I. Zink, F. Tamanoi, Biocompatibility, biodistribution, and drug-delivery efficiency of mesoporous silica nanoparticles for cancer therapy in animals, Small 6 (16) (2010) 1794–1805.

[4] J. Lu, Z. Li, J.I. Zink, F. Tamanoi, In vivo tumor suppression efficacy of mesoporous silica nanoparticles-based drug-delivery system: enhanced efficacy by folate modification, Nanomedicine 8 (2) (2012) 212–220.

[5] X. Liu, P. Lin, I. Perrett, J. Lin, Y.-P. Liao, C.H. Chang, J. Jiang, N. Wu, T. Donahue, Z. Wainberg, A.E. Nel, H. Meng, Tumor-penetrating peptide enhances transcytosis of silicasome-based chemotherapy for pancreatic cancer, J. Clin. Invest. 127 (5) (2017) 2007–2018.

[6] F. Chen, H. Hong, S. Shi, S. Goel, H.F. Valdovinos, R. Hernandez, C.P. Theuer, T.E. Barnhart, W. Cai, Engineering of hollow mesoporous silica nanoparticles for remarkably enhanced tumor active targeting efficacy, Sci. Rep. 4 (2014) 5080.

[7] J. Zhou, M. Li, W.Q. Lim, Z. Luo, S.Z.F. Phua, R. Huo, L. Li, K. Li, L. Dai, J. Liu, K. Cai, Y. Zhao, A transferrin-conjugated hollow nanoplatform for redox-controlled and targeted chemotherapy of tumor with reduced inflammatory reactions, Theranostics 8 (1) (2018) 518–532.

[8] Z. Luo, X. Ding, Y. Hu, S. Wu, Y. Xiang, Y. Zeng, B. Zhang, H. Yan, H. Zhang, L. Zhu, J. Liu, J. Li, K. Cai, Y. Zhao, Engineering a hollow nanocontainer platform with multifunctional molecular machines for tumor-targeted therapy in vitro and in vivo, ACS Nano 7 (11) (2013) 10271–10284.

[9] J. Finlay, C.M. Roberts, J. Dong, J.I. Zink, F. Tamanoi, C.A. Glackin, Mesoporous silica nanoparticle delivery of chemically modified siRNA against TWIST1 leads to reduced tumor burden, Nanomedicine 11 (7) (2015) 1657–1666.

[10] C.M. Roberts, S.A. Shahin, W. Wen, J.B. Finlay, J. Dong, R. Wang, T.H. Dellinger, J.I. Zink, F. Tamanoi, C.A. Glackin, Nanoparticle delivery of siRNA against TWIST to reduce drug resistance and tumor growth in ovarian cancer models, Nanomedicine 13 (3) (2017) 965–976.

[11] S.A. Shahin, R. Wang, S.I. Simargi, A. Contreras, L. Parra Echavarria, L. Qu, W. Wen, T. Dellinger, J. Unternaehrer, F. Tamanoi, J.I. Zink, C.A. Glackin, Hyaluronic acid conjugated nanoparticle delivery of siRNA against TWIST reduces tumor burden and enhances sensitivity to cisplatin in ovarian cancer, Nanomedicine 14 (4) (2018) 1381–1394.

[12] L. Maggini, I. Cabrera, A. Ruiz-Carretero, E.A. Prasetyanto, E. Robinet, L. De Cola, Breakable mesoporous silica nanoparticles for targeted drug delivery, Nanoscale 8 (13) (2016) 7240–7247.

[13] J. Croissant, X. Cattoën, M.W.C. Man, A. Gallud, L. Raehm, P. Trens, M. Maynadier, J.-O. Durand, Biodegradable ethylene-bis(propyl)disulfide-based periodic mesoporous

organosilica nanorods and nanospheres for efficient in-vitro drug delivery, Adv. Mater. 26 (35) (2014) 6174–6180.
[14] J.G. Croissant, Y. Fatieiev, K. Julfakyan, J. Lu, A.-H. Emwas, D.H. Anjum, H. Omar, F. Tamanoi, J.I. Zink, N.M. Khashab, Biodegradable oxamide-phenylene-based mesoporous organosilica nanoparticles with unprecedented drug payloads for delivery in cells, Chemistry 22 (42) (2016) 14806–14811.
[15] B.T. Vu, S.A. Shahin, J. Croissant, Y. Fatieiev, K. Matsumoto, T. Le-Hoang Doan, T. Yik, S. Simargi, A. Conteras, L. Ratliff, C.M. Jimenez, L. Raehm, N. Khashab, J.-O. Durand, C. Glackin, F. Tamanoi, Chick chorioallantoic membrane assay as an in vivo model to study the effect of nanoparticle-based anticancer drugs in ovarian cancer, Sci. Rep. 8 (1) (2018) 8524.
[16] K.S. Butler, P.N. Durfee, C. Theron, C.E. Ashley, E.C. Carnes, C.J. Brinker, Protocells: modular mesoporous silica nanoparticle-supported lipid bilayers for drug delivery, Small 12 (16) (2016) 2173–2185.
[17] J. Liu, A. Stace-Naughton, X. Jiang, C.J. Brinker, Porous nanoparticle supported lipid bilayers (protocells) as delivery vehicles, J. Am. Chem. Soc. 131 (4) (2009) 1354–1355.
[18] J. Lu, X. Liu, Y.-P. Liao, F. Salazar, B. Sun, W. Jiang, C.H. Chang, J. Jiang, X. Wang, A.M. Wu, H. Meng, A.E. Nel, Nano-enabled pancreas cancer immunotherapy using immunogenic cell death and reversing immunosuppression, Nat. Commun. 8 (1) (2017) 1811.
[19] X. Liu, A. Situ, Y. Kang, K.R. Villabroza, Y. Liao, C.H. Chang, T. Donahue, A.E. Nel, H. Meng, Irinotecan delivery by lipid-coated mesoporous silica nanoparticles shows improved efficacy and safety over liposomes for pancreatic cancer, ACS Nano 10 (2) (2016) 2702–2715.
[20] C. Hom, J. Lu, M. Liong, H. Luo, Z. Li, J.I. Zink, F. Tamanoi, Mesoporous silica nanoparticles facilitate delivery of siRNA to shutdown signaling pathways in mammalian cells, Small 6 (11) (2010) 1185–1190.
[21] T. Xia, M. Kovochich, M. Liong, H. Meng, S. Kabehie, S. George, J.I. Zink, A.E. Nel, Polyethyleneimine coating enhances the cellular uptake of mesoporous silica nanoparticles and allows safe delivery of siRNA and DNA constructs, ACS Nano 3 (10) (2009) 3273–3286.
[22] H. Meng, M. Xue, T. Xia, Z. Ji, D.Y. Tarn, J.I. Zink, A.E. Nel, Use of size and a copolymer design feature to improve the biodistribution and the enhanced permeability and retention effect of doxorubicin-loaded mesoporous silica nanoparticles in a murine xenograft tumor model, ACS Nano 5 (5) (2011) 4131–4144.
[23] H. Meng, W.X. Mai, H. Zhang, M. Xue, T. Xia, S. Lin, X. Wang, Y. Zhao, Z. Ji, J.I. Zink, A.E. Nel, Codelivery of an optimal drug/siRNA combination using mesoporous silica nanoparticles to overcome drug resistance in breast cancer in vitro and in vivo, ACS Nano 7 (2) (2013) 994–1005.
[24] H. Meng, Y. Zhao, J. Dong, M. Xue, Y.-S. Lin, Z. Ji, W.X. Mai, H. Zhang, C.H. Chang, C.J. Brinker, J.I. Zink, A.E. Nel, Two-wave nanotherapy to target the stroma and optimize gemcitabine delivery to a human pancreatic cancer model in mice, ACS Nano 7 (11) (2013) 10048–10065.
[25] M. Xie, Y. Xu, H. Shen, S. Shen, Y. Ge, J. Xie, Negative-charge-functionalized mesoporous silica nanoparticles as drug vehicles targeting hepatocellular carcinoma, Int. J. Pharm. 474 (1) (2014) 223–231.
[26] X. Ma, Y. Zhao, K.W. Ng, Y. Zhao, Integrated hollow mesoporous silica nanoparticles for target drug/siRNA co-delivery, Chemistry 19 (46) (2013) 15593–15603.
[27] J.-F. Chen, H.-M. Ding, J.-X. Wang, L. Shao, Preparation and characterization of porous hollow silica nanoparticles for drug delivery application, Biomaterials 25 (4) (2004) 723–727.

[28] B.K. Seon, A. Haba, F. Matsuno, N. Takahashi, M. Tsujie, X. She, N. Harada, S. - Uneda, T. Tsujie, H. Toi, H. Tsai, Y. Haruta, Endoglin-targeted cancer therapy, Curr. Drug Deliv. 8 (1) (2011) 135–143.
[29] L. Li, T. Liu, C. Fu, X. Meng, H. Liu, Size effect of mesoporous and hollow silica nanoparticles on solid tumor targeting and penetration, J. Nanosci. Nanotechnol. 16 (7) (2016) 6766–6772.
[30] W. Cheng, C. Liang, X. Wang, H.-i. Tsai, G. Liu, Y. Peng, J. Nie, L. Huang, L. Mei, X. Zeng, A drug-self-gated and tumor microenvironment-responsive mesoporous silica vehicle: "four-in-one" versatile nanomedicine for targeted multidrug-resistant cancer therapy, Nanoscale 9 (43) (2017) 17063–17073.
[31] R. Hernandez, H.-R. Tseng, J.W. Wong, J.F. Stoddart, J.I. Zink, An operational supramolecular nanovalve, J. Am. Chem. Soc. 126 (11) (2004) 3370–3371.
[32] N. Song, Y.-W. Yang, Molecular and supramolecular switches on mesoporous silica nanoparticles, Chem. Soc. Rev. 44 (11) (2015) 3474–3504.
[33] Z. Xu, S. Liu, Y. Kang, M. Wang, Glutathione- and pH-responsive nonporous silica prodrug nanoparticles for controlled release and cancer therapy, Nanoscale 7 (13) (2015) 5859–5868.
[34] C.-H. Lee, S.-H. Cheng, I.-P. Huang, J.S. Souris, C.-S. Yang, C.-Y. Mou, L.-W. Lo, Intracellular pH-responsive mesoporous silica nanoparticles for the controlled release of anticancer chemotherapeutics, Angew. Chem. Int. Ed. 49 (44) (2010) 8214–8219.
[35] Y.-L. Zhao, Z. Li, S. Kabehie, Y.Y. Botros, J.F. Stoddart, J.I. Zink, pH-operated nanopistons on the surfaces of mesoporous silica nanoparticles, J. Am. Chem. Soc. 132 (37) (2010) 13016–13025.
[36] Q. Gan, X. Lu, Y. Yuan, J. Qian, H. Zhou, X. Lu, J. Shi, C. Liu, A magnetic, reversible pH-responsive nanogated ensemble based on Fe_3O_4 nanoparticles-capped mesoporous silica, Biomaterials 32 (7) (2011) 1932–1942.
[37] E. Aznar, M.D. Marcos, R. Martínez-Máñez, F. Sancenón, J. Soto, P. Amorós, C. Guillem, pH- and photo-switched release of guest molecules from mesoporous silica supports, J. Am. Chem. Soc. 131 (19) (2009) 6833–6843.
[38] H. Meng, M. Xue, T. Xia, Y.-L. Zhao, F. Tamanoi, J.F. Stoddart, J.I. Zink, A.E. Nel, Autonomous in vitro anticancer drug release from mesoporous silica nanoparticles by pH-sensitive nanovalves, J. Am. Chem. Soc. 132 (36) (2010) 12690–12697.
[39] F. Chen, Y. Zhu, Chitosan enclosed mesoporous silica nanoparticles as drug nanocarriers: sensitive response to the narrow pH range, Microporous Mesoporous Mater. 150 (2012) 83–89.
[40] H. Kim, S. Kim, C. Park, H. Lee, H.J. Park, C. Kim, Glutathione-induced intracellular release of guests from mesoporous silica nanocontainers with cyclodextrin gatekeepers, Adv. Mater. 22 (38) (2010) 4280–4283.
[41] R. Liu, X. Zhao, T. Wu, P. Feng, Tunable redox-responsive hybrid nanogated ensembles, J. Am. Chem. Soc. 130 (44) (2008) 14418–14419.
[42] L. Zhong, C. Kaiyong, H. Yan, Z. Li, L. Peng, D. Lin, Y. Weihu, Mesoporous silica nanoparticles end-capped with collagen: redox-responsive nanoreservoirs for targeted drug delivery, Angew. Chem. Int. Ed. 50 (3) (2011) 640–643.
[43] Y. Cui, H. Dong, X. Cai, D. Wang, Y. Li, Mesoporous silica nanoparticles capped with disulfide-linked PEG gatekeepers for glutathione-mediated controlled release, ACS Appl. Mater. Interfaces 4 (6) (2012) 3177–3183.
[44] Q. Zhang, F. Liu, K.T. Nguyen, X. Ma, X. Wang, B. Xing, Y. Zhao, Multifunctional mesoporous silica nanoparticles for cancer-targeted and controlled drug delivery, Adv. Funct. Mater. 22 (24) (2012) 5144–5156.
[45] C.-L. Zhu, X.-Y. Song, W.-H. Zhou, H.-H. Yang, Y.-H. Wen, X.-R. Wang, An efficient cell-targeting and intracellular controlled-release drug delivery system based on MSN-PEM-aptamer conjugates, J. Mater. Chem. 19 (41) (2009) 7765–7770.

[46] A. Bernardos, L. Mondragón, E. Aznar, M.D. Marcos, R. Martínez-Máñez, F. Sancenón, J. Soto, J.M. Barat, E. Pérez-Payá, C. Guillem, P. Amorós, Enzyme-responsive intracellular controlled release using nanometric silica mesoporous supports capped with "Saccharides", ACS Nano 4 (11) (2010) 6353–6368.
[47] L. Mondragón, N. Mas, V. Ferragud, C. de la Torre, A. Agostini, R. Martínez-Máñez, F. Sancenón, P. Amorós, E. Pérez-Payá, M. Orzáez, Enzyme-responsive intracellular-controlled release using silica mesoporous nanoparticles capped with ε-poly-L-lysine, Chemistry 20 (18) (2014) 5271–5281.
[48] J. Liu, B. Zhang, Z. Luo, X. Ding, J. Li, L. Dai, J. Zhou, X. Zhao, J. Ye, K. Cai, Enzyme responsive mesoporous silica nanoparticles for targeted tumor therapy in vitro and in vivo, Nanoscale 7 (8) (2015) 3614–3626.
[49] X. He, Y. Zhao, D. He, K. Wang, F. Xu, J. Tang, ATP-responsive controlled release system using aptamer-functionalized mesoporous silica nanoparticles, Langmuir 28 (35) (2012) 12909–12915.
[50] E. Aznar, C. Coll, M.D. Marcos, R. Martínez-Máñez, F. Sancenón, J. Soto, P. Amorós, J. Cano, E. Ruiz, Borate-driven gatelike scaffolding using mesoporous materials functionalised with saccharides, Chemistry 15 (28) (2009) 6877–6888.
[51] E. Climent, A. Bernardos, R. Martínez-Máñez, A. Maquieira, M.D. Marcos, N. Pastor-Navarro, R. Puchades, F. Sancenón, J. Soto, P. Amorós, Controlled delivery systems using antibody-capped mesoporous nanocontainers, J. Am. Chem. Soc. 131 (39) (2009) 14075–14080.
[52] D.P. Ferris, Y.-L. Zhao, N.M. Khashab, H.A. Khatib, J.F. Stoddart, J.I. Zink, Light-operated mechanized nanoparticles, J. Am. Chem. Soc. 131 (5) (2009) 1686–1688.
[53] Q. Yuan, Y. Zhang, T. Chen, D. Lu, Z. Zhao, X. Zhang, Z. Li, C.-H. Yan, W. Tan, Photon-manipulated drug release from a mesoporous nanocontainer controlled by azobenzene-modified nucleic acid, ACS Nano 6 (7) (2012) 6337–6344.
[54] D. Juyao, J.I. Zink, Light or heat? The origin of cargo release from nanoimpeller particles containing upconversion nanocrystals under IR irradiation, Small 11 (33) (2015) 4165–4172.
[55] L. Jie, C. Eunshil, T. Fuyuhiko, J.I. Zink, Light-activated nanoimpeller-controlled drug release in cancer cells, Small 4 (4) (2008) 421–426.
[56] T.M. Guardado-Alvarez, L. Sudha Devi, M.M. Russell, B.J. Schwartz, J.I. Zink, Activation of snap-top capped mesoporous silica nanocontainers using two near-infrared photons, J. Am. Chem. Soc. 135 (38) (2013) 14000–14003.
[57] M. Gary-Bobo, Y. Mir, C. Rouxel, D. Brevet, O. Hocine, M. Maynadier, A. Gallud, A. Da Silva, O. Mongin, M. Blanchard-Desce, S. Richeter, B. Loock, P. Maillard, A. Morère, M. Garcia, L. Raehm, J.-O. Durand, Multifunctionalized mesoporous silica nanoparticles for the in vitro treatment of retinoblastoma: drug delivery, one and two-photon photodynamic therapy, Int. J. Pharm. 432 (1) (2012) 99–104.
[58] P. Saint-Cricq, S. Deshayes, J.I. Zink, A.M. Kasko, Magnetic field activated drug delivery using thermodegradable azo-functionalised PEG-coated core-shell mesoporous silica nanoparticles, Nanoscale 7 (31) (2015) 13168–13172.
[59] P.-J. Chen, S.-H. Hu, C.-S. Hsiao, Y.-Y. Chen, D.-M. Liu, S.-Y. Chen, Multifunctional magnetically removable nanogated lids of Fe_3O_4-capped mesoporous silica nanoparticles for intracellular controlled release and MR imaging, J. Mater. Chem. 21 (8) (2011) 2535–2543.
[60] J.E. Lee, N. Lee, H. Kim, J. Kim, S.H. Choi, J.H. Kim, T. Kim, I.C. Song, S.P. Park, W.K. Moon, T. Hyeon, Uniform mesoporous dye-doped silica nanoparticles decorated with multiple magnetite nanocrystals for simultaneous enhanced magnetic resonance imaging, fluorescence imaging, and drug delivery, J. Am. Chem. Soc. 132 (2) (2010) 552–557.

[61] Y. Chen, H. Chen, D. Zeng, Y. Tian, F. Chen, J. Feng, J. Shi, Core/shell structured hollow mesoporous nanocapsules: a potential platform for simultaneous cell imaging and anticancer drug delivery, ACS Nano 4 (10) (2010) 6001–6013.
[62] A. Baeza, E. Guisasola, E. Ruiz-Hernández, M. Vallet-Regí, Magnetically triggered multidrug release by hybrid mesoporous silica nanoparticles, Chem. Mater. 24 (3) (2012) 517–524.
[63] C.R. Thomas, D.P. Ferris, J.-H. Lee, E. Choi, M.H. Cho, E.S. Kim, J.F. Stoddart, J.-S. Shin, J. Cheon, J.I. Zink, Noninvasive remote-controlled release of drug molecules in vitro using magnetic actuation of mechanized nanoparticles, J. Am. Chem. Soc. 132 (31) (2010) 10623–10625.
[64] A.A. Hwang, J. Lu, F. Tamanoi, J.I. Zink, Functional nanovalves on protein-coated nanoparticles for in vitro and in vivo controlled drug delivery, Small 11 (3) (2015) 319–328.
[65] L. Dai, Q. Zhang, H. Gu, K. Cai, Facile synthesis of yolk-shell silica nanoparticles for targeted tumor therapy, J. Mater. Chem. B 3 (42) (2015) 8303–8313.
[66] S.R. Gayam, P. Venkatesan, Y.-M. Sung, S.-Y. Sung, S.-H. Hu, H.-Y. Hsu, S.-P. Wu, An NAD(P)H:quinone oxidoreductase 1 (NQO1) enzyme responsive nanocarrier based on mesoporous silica nanoparticles for tumor targeted drug delivery in vitro and in vivo, Nanoscale 8 (24) (2016) 12307–12317.

CHAPTER FIVE

Nanoparticle Delivery of TWIST Small Interfering RNA and Anticancer Drugs: A Therapeutic Approach for Combating Cancer

Carlotta A. Glackin[1]
Developmental and Stem Cell Biology, City of Hope Medical Center, Duarte, CA, United States
[1]Corresponding author: e-mail address: cglackin@coh.org

Contents

Abstract

Breast and ovarian cancer are the leading cause of cancer-related deaths in women in the United States with over 232,000 new Breast Cancer (BC) diagnoses expected in 2018 and almost 40,000 deaths and an estimated 239,000 new ovarian cancer (OC) cases and 152,000 deaths worldwide annually. OC is the most lethal gynecologic malignancy. This high mortality rate is due to tumor recurrence and metastasis, primarily caused by chemoresistant cancer stem-like cells (CSCs). Triple Negative Breast Cancer (TNBC) patients also become resistant to chemotherapy due to recurrence of CSCs. Currently, no ovarian or breast cancer therapies target CSC specifically. *TWIST* is overexpressed in the majority of chemoresistant cancers resulting in a low survival rate. Our long-term goal is to develop novel treatments for women with ovarian and breast cancer, specifically treatments that sensitize chemoresistant tumors.

The Enzymes, Volume 44
ISSN 1874-6047
https://doi.org/10.1016/bs.enz.2018.08.004
© 2018 Elsevier Inc.
All rights reserved.

Despite successful initial surgery and chemotherapy, over 70% of advanced EOC will recur, and only 15–30% of recurrent disease will respond to chemotherapy (Cortez et al., 2017; Berezhnaya, 2010; Jackson et al., 2015). Moreover, drug resistance causes treatment failure in over 90% of patients with metastatic disease (Solmaz et al., 2015). Thus, recurrent metastatic disease is a major clinical challenge without effective therapy.

One of the major challenges in the treatment of breast cancer is the presence of a subpopulation of cancer cells that are chemoresistant (CRC) and metastatic. Given that metastasis is the driving force behind mortality for breast and ovarian cancer patients, it is essential to identify the characteristics of these aberrant cancer cells that allow them to spread to distant sites in the body and develop into metastatic tumors. Understanding the metastatic mechanisms driving cancer cell dispersal will open the door to developing novel therapies that prevent metastasis and improve long-term outcomes for patients. In this chapter we assess the feasibility of targeting the Twist and EMT signaling pathways in breast and ovarian cancer. Additional discussions of the pathways that mediate epithelial-mesenchymal transition (EMT), a process that can give rise to chemoresistance. We review potential treatment strategies for targeting EMT and drug resistance as well as the problems that may arise with these targeted delivery therapeutic approaches. Finally, we examine recent advances in the field, including cancer stem cell targeted nanoparticle delivery and small interference RNA (siRNA) technology, and discuss the impact that these approaches may have on translating much needed therapeutic approaches into the clinic, for the benefit of patients battling this devastating disease.

1. INTRODUCTION

Ovarian cancer accounts for an estimated 239,000 new cases and 152,000 deaths worldwide annually [1]. This disease has both the highest morbidity and mortality rates among cancers of the reproductive system. Each year, at least 20,000 women in the United States are diagnosed with ovarian cancer [1,2]. Recurrent epithelial ovarian cancer (EOC) is almost uniformly lethal. Tumor recurrence and metastases are primarily responsible for the 70% 5-year mortality of advanced EOC. Despite successful initial surgery and chemotherapy, over 70% of advanced EOC will recur, and only 15–30% of recurrent disease will respond to chemotherapy [1–3]. Moreover, drug resistance causes treatment failure in over 90% of patients with metastatic disease [4]. Thus, recurrent metastatic disease is a major clinical challenge without effective therapy.

Chemoresistant cancer stem cells (CSCs) are the primary cause of relapse and metastasis. Activation of several mechanisms in CSCs, including drug efflux or inactivation, signaling pathways that enhance survival, and alter DNA damage response, can lead to chemoresistance [5]. Currently, no therapeutics specifically target CSC. These cancer-related events include cancer invasion, metastasis, and resistance to apoptotic signals, including chemotherapy and immunotherapy, and the acquisition of stem cell-like properties. There is increasing evidence suggesting a relationship between CSC and drug resistance, and there is a need to elucidate how these processes correlate and how chemotherapy may accelerate these changes [6]. Despite up to a 90% response rate to initial therapy, once tumor cells become drug resistant, no standard second-line therapy exists that can eliminate the tumor cells that have been disseminated throughout the peritoneal cavity. What is needed is a novel therapeutic strategy that can address the parallel concerns of metastasis and drug resistance. Ideally, a novel treatment would target a single protein or signaling cascade that influences both processes. One potential target for such therapy is the transcription factor TWIST, involved in both metastasis and drug resistance. Breast cancer is the leading cause of cancer-related deaths in the United States with over 232,000 new diagnoses expected in 2014 and almost 40,000 deaths [7,8]. The "triple-negative breast cancer" (TNBC) (estrogen receptor [ER]-negative, progesterone [PR]-negative, and human epidermal growth factor receptor 2 [HER2]-negative) subtype, as well as a subtype called "claudin-low" based on gene expression profiling, pose a particularly difficult challenge because they are resistant to chemotherapy and have a high incidence of metastasis, and patients with tumors have a poor prognosis [9]. There is thus a critical need for novel efficacious therapies for patients with claudin-low, TNBC. TNBC, claudin-low type tumors, is characterized by high expression levels of epithelial to mesenchymal (EMT) markers such as *TWIST*. Targeting EMT by inhibiting *TWIST* using siRNA in TNBC, claudin-low cells may therefore sensitize them to further drug treatments.

In this chapter we assess the feasibility of targeting the Twist and EMT signaling pathways in BC and OC. These pathways mediate epithelial-mesenchymal transition (EMT), a process that can give rise to metastatic and chemoresistant cells (CRCs). We review potential treatment strategies for targeting EMT and drug resistance as well as the problems that may arise with therapeutic delivery approaches in breast cancer models. Finally, we examine the most recent advances in the field, including nanoparticle delivery and siRNA technology, and discuss the impact that these approaches may have on translating much needed therapeutic approaches into the clinic, for the benefit of patients battling MBC.

2. EPITHELIAL TO MESENCHYMAL TRANSITION

Epithelial-mesenchymal transition is a biological process that allows a polarized epithelial cell, which normally interacts with basement membrane via its basal surface and with surrounding cells, to undergo multiple biochemical changes that enable it to assume a mesenchymal cell phenotype. This phenotype includes enhanced migratory capacity, invasiveness, elevated resistance to apoptosis, and greatly increased production of extracellular matrix (ECM) components [10]. The completion of an EMT event is signaled by the degradation of underlying basement membrane and the formation of a mesenchymal stem cell that can migrate away from the epithelial layer from which it originated [11]. Epithelial-mesenchymal transition is a key mechanism for the spread of cancer cells which undergo changes that enable them to detach from surrounding cells and become mobile [12–15]. Cells that have undergone EMT are particularly prevalent in patients with BC and OC. Thus, approaches that preferentially target these cells may lead to the development of novel and more efficacious therapies than are currently available.

3. TWIST: AN ATTRACTIVE THERAPEUTIC TARGET

The Twist family of basic helix-loop-helix transcription factors is involved in EMT [16], and Twist is a known activator of EMT in cancer cells [17,18]. It functions by activating several target genes that promote cellular de-differentiation and cell mobility. In addition, Twist1 is thought to promote the cancer stem cell phenotype [11], inhibit apoptosis [19,20], and contribute to chemotherapy resistance [21,22]. Overexpression of Twist or methylation of its promoter is common in metastatic carcinomas [23,24], including in aggressive and metastatic forms of breast cancer [25,26].

We and others have shown that Twist1 and Twist2 transcription factors mediate EMT target genes involved in cell migration/invasion [27], cancer stem cell self-renewal [11,28], multi-drug resistance [29–32], apoptosis [33], and immune surveillance [34] (Fig. 1). The central role of Twist as a master regulator of EMT in breast epithelial cells thus supports researching it as a promising "druggable" target for CRC therapy. Twist1 is also an attractive therapeutic target for MBC because it is rarely expressed in normal adult tissues [35]. Therefore, systemic administration of a Twist1 inhibitor could have a significant impact on Twist1-overexpressing cancer cells with minimal

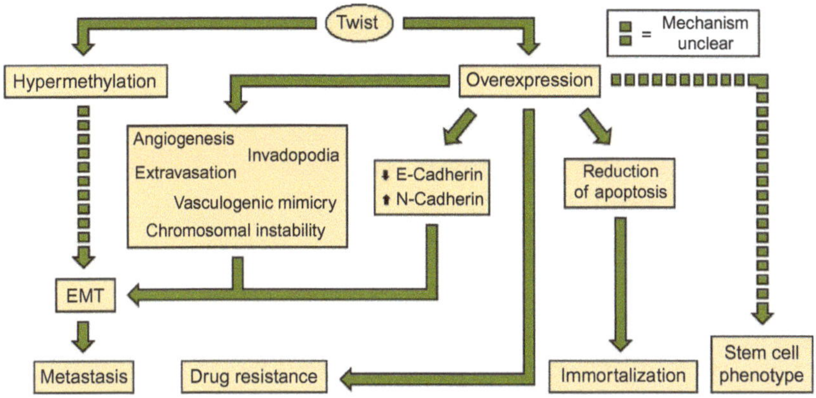

Fig. 1 The central role of Twist as a master regulator of EMT.

side effects in other tissues. The inactivation of Twist by small RNA technology (siRNA) or chemotherapeutic approaches has proved successful [36–39], and several inhibitors that are antagonistic to the upstream or downstream molecules of Twist signaling have also been identified [40].

4. TARGETING EMT AND CSC SIGNALING: AN ATTRACTIVE THERAPEUTIC APPROACH

TWIST, a protein and master regulator of EMT, delivered by nanoparticle carriers targeting cancer stem cells (CSCs) to reverse chemoresistance and when combined with standard therapy improves survival in drug-resistant ovarian and breast cancer patients. TWIST is a basic helix-loop-helix transcription factor critical to organismal development. Twist protein is often aberrantly activated in many cancers, where it is expressed in CSCs and again drives an epithelial to mesenchymal transition (EMT) event. EMT allows cell layers to invade and form mesodermal structures. TWIST expressing cancer cells are able to migrate away from their primary site and metastasize. Recent work reveals that CSCs also tend to be resistant to chemotherapy. CSCs with this phenotype are of great concern in epithelial ovarian cancer (EOC), one of the most lethal gynecologic cancers. Besides Twist, other factors such as Snail and Let-7 contribute to the CSC phenotype in ovarian cancer and regulate PARP1 [41]. PARP1 inhibitors (PARPi) have been approved for treatment of certain ovarian cancers [42,43], with recent trials demonstrating significant increases in survival [42,44–46]. Targeted therapies such as PARPi are a cornerstone of precision medicine, but resistance will remain a critical barrier [47] until CSC can be eliminated [48]. The mechanism of action of PARPi could

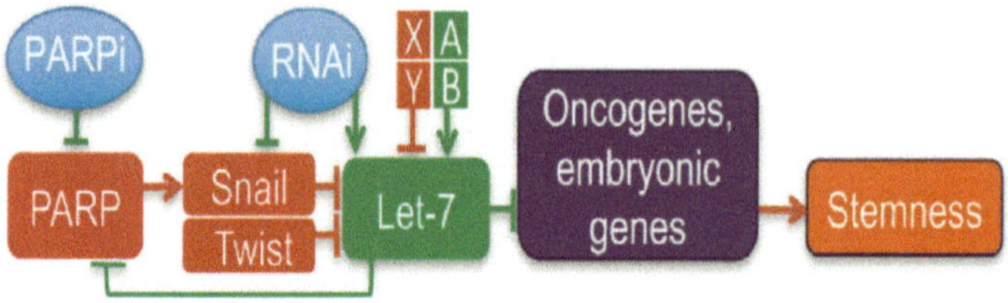

Fig. 2 PARP1 inhibitors (PARPi), and inhibition of the Snail/let-7 axis, cooperate to block stemness, inhibit invasion, and cause apoptosis.

extend beyond PARP's most famous role, DNA damage response (DDR). In addition to PARP1's function in DDR, it increases Snail activity [49–51], and plays a role in establishing the stem cell phenotype [52–55]. Further, let-7 inhibits PARP1 [56]. Thus, PARP1 inhibitors (PARPi), and inhibition of the Snail/let-7 axis, could cooperate to block stemness, inhibit invasion, and cause apoptosis (Fig. 2) [57].

5. SMALL INTERFERING RNAs: FROM BASIC SCIENCE TO CLINICAL APPLICATIONS

RNA interference (RNAi) using double-stranded RNA was first demonstrated in *C. elegans*. A double-stranded RNA molecule produced a greater effect than either single strand alone [58]. Following this discovery, vital elements of the RNAi mechanism were quickly discovered in flies, plants, and eventually mammals [59]. In mammals, 21-nucleotide long RNA fragments called small interfering RNAs (siRNAs) are the effectors of RNAi [60]. These fragments bind to a target mRNA, and then form the RNA induced silencing complex (RISC) [61,62]. RISC also contains Argonaute proteins, specifically Ago2, that cleave the mRNA thus preventing protein translation [63].

In recent years, efforts have begun to use siRNA as a therapeutic agent. Preclinical and clinical trials have demonstrated that siRNA can be delivered to cells *in vivo* and retain its function, knocking down target genes in both prostate cancer and melanoma [64,65].

5.1 Concerns With siRNA Therapy

A drawback to RNAi-based therapy is that RNA tends to be susceptible to nuclease degradation, and is thus unstable in the body. For it to reach its intended target intact and without activating unwanted immune responses, chemical modification of the siRNA is often needed. A wide variety of modifications have been described, but there are several straightforward modifications that can greatly increase siRNA efficacy. The first concern is the sequence of the siRNA; certain sequences are potent activators of toll-like receptors (TLRs), such as TLR-7, TLR-8, and TLR-9, which are key immune response proteins evolved for the detection of pathogenic nucleic acids [66]. Furthermore, alternative regulators such as RIG1 respond to siRNA sequences independent of TLRs [67]. In addition to avoiding immunogenic motifs, modifications to the backbone of the siRNA are required to overcome the additional challenges of siRNA instability. Substitutions on the 2′ carbon of ribose afford protection from nuclease degradation [68]. A common modification, 2′ *O*-methyl, also assists in blocking immune system activation by siRNA [68]. A less common but multifunctional modification is the inverted abasic ribose. Placed at the ends of an siRNA strand, it prevents the nuclease degradation of the entire duplex [68]. Moreover, the abnormal structure at the ends of the strand greatly lowers the efficiency of RISC loading. Modifying a passenger strand in this fashion, but leaving the guide strand unmodified, favors the incorporation of the guide strand into RISC, thus increasing the efficiency and potency of the siRNA [69,70]. Many other modifications exist, but our previous work suggests that these few are sufficient to promote siRNA efficacy in an *in vivo* environment [71].

A complementary method for protecting siRNA is to complex it with a delivery vehicle, and numerous options have been described in the literature. Many formulations, including some currently in clinical trials, rely on lipid-based encapsulation of the siRNA molecules [72,73]. Others have relied on peptide-based targeting of RNA to cells or linking to polymer nanoparticles [74].

6. POLYAMIDOAMINE DENDRIMERS: NANOPARTICLES FOR DELIVERING siRNA TO TUMOR CELLS

One form of nanoparticle-mediated delivery is the use of polyamidoamine (PAMAM) dendrimers as a vehicle. These branched polymers electrostatically interact with RNA, and siRNA that adheres to these particles can knockdown target genes after delivery to prostate cancer cells [75,76]. In addition, a trial at City of Hope has shown efficacy of siRNA linked to PAMAM dendrimers in a model of HIV infection [77]. PAMAM dendrimers are not targeted to any particular cell type, but HIV viral proteins are not present except in infected cells. This technique is ideal for adaptation to targeting *TWIST*, which likewise is rarely expressed outside the intended target cells.

Design and optimization of PAMAM dendrimers and their association with siRNA are a key concern for proper knockdown, with some researchers suggesting that a larger dendrimer carrying a larger payload is more stable [78]. Generation 5 dendrimers (those possessing five layers of branch points, for 96 terminal amino groups) can carry three siRNA duplexes. As seen for generation 6 dendrimers, although no specific binding pocket is designed as part of these dendrimers, the siRNA partially fits into the gaps between amino groups, but mostly lies across the surface of the particle [79] (Fig. 3A, [80]). However, dendrimer size tends to correlate with cytotoxicity—the higher the generation number, the greater the toxicity, which has been shown to

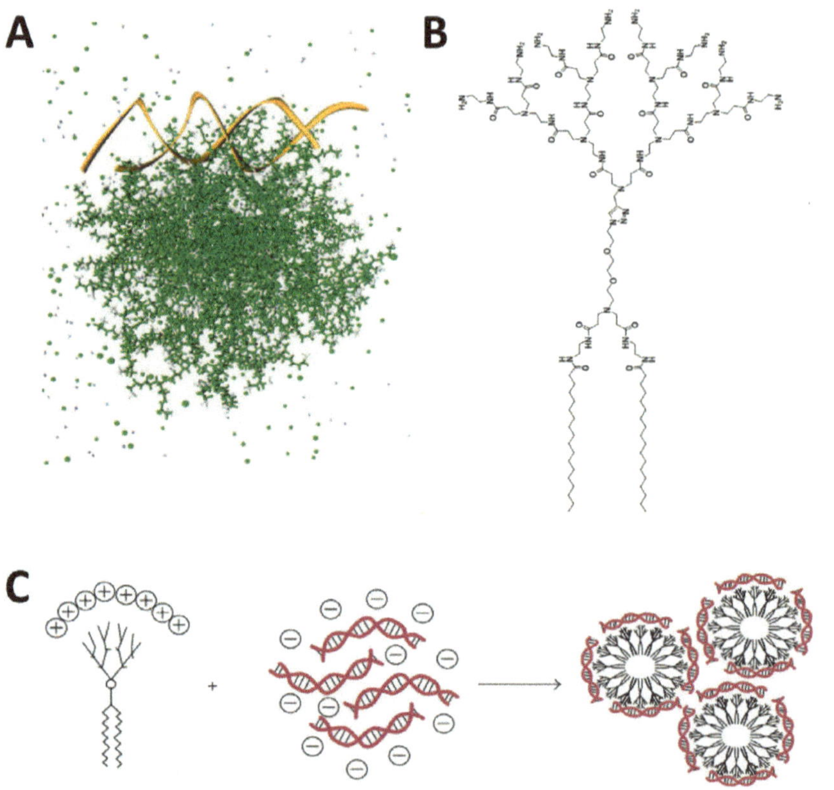

Fig. 3 (A) 3D representation of a generation 6 dendrimer (green) with associated siRNA (gold) [80]. (B) An amphiphilic generation 3 dendrimer [81]. (C) Dendrimers form micelles around hydrophobic cores upon exposure to negatively charged siRNA [82].

be a result of increased reactive oxygen species (ROS) in target cells [83,84]. *In vivo*, mitochondrial activity in the brain can be disrupted by these ROS [85]. There are several ways to reduce dendrimer toxicity, including coating the particles with polyethylene glycol or changing the terminal amines to alternative functional groups [84,86].

Using smaller dendrimers may also solve the toxicity problem [86]. Yu et al. demonstrated effective siRNA delivery and gene knockdown with a small generation 3 dendrimer with partial hydrophobic character, representing an alternative school of design emphasizing delivery over carrying capacity [87] (Fig. 3B, [81]). It was later revealed that amphiphilic dendrimers form micelle structures in aqueous solution, which repack into smaller micelles coated in the siRNA once siRNA is added to the solution [88] (Fig. 3C, [82]). We have previously used these YTZ3-15 dendrimers to knockdown *TWIST1 in vitro*, which produced a drop in migratory potential. In addition, dendrimers carrying siRNA exhibited short-term tumor tropism when injected into tumor bearing mice [82]. Chapter "Drug delivery and bone infection" by Vallet-Regí et al. discusses additional studies utilizing YTZ3-15.

7. MESOPOROUS SILICA NANOPARTICLES: MULTIFUNCTIONAL DRUG AND siRNA CARRIERS

An alternative nanoparticle platform is the mesoporous silica nanoparticle (MSN). MSNs are spheroids composed of silica and containing pores formed by the removal of surfactant during synthesis, and are tunable from 50 to 200 nm diameter [89,90] (Fig. 4, [82]). MSNs can be modified to carry a number of targeting moieties on their surface, but exhibit a degree of inherent tumor tropism thanks to the enhanced permeability and retention (EPR) effect [91]. In EPR, particles enter the tumor microenvironment due to the leakier vasculature rapidly produced to support the growing tumor [91]. Due to poor lymphatic drainage and uptake by tumor cells, nanoparticles such as MSNs tend to accumulate at tumors compared to the body as a whole [92]. MSNs can carry a variety of drugs within their pore structure, but can also be modified to carry siRNA. Low molecular weight polyethyleneimine (PEI) can protect and carry siRNA or DNA into cells, and prior work has demonstrated that PEI-modified MSNs are successful at gene silencing in human cells [90,93]. PEI coating can increase the uptake of MSNs into cells, but maintaining low PEI molecular weight is important for effective delivery of siRNA and to avoid cytotoxicity [93,94].

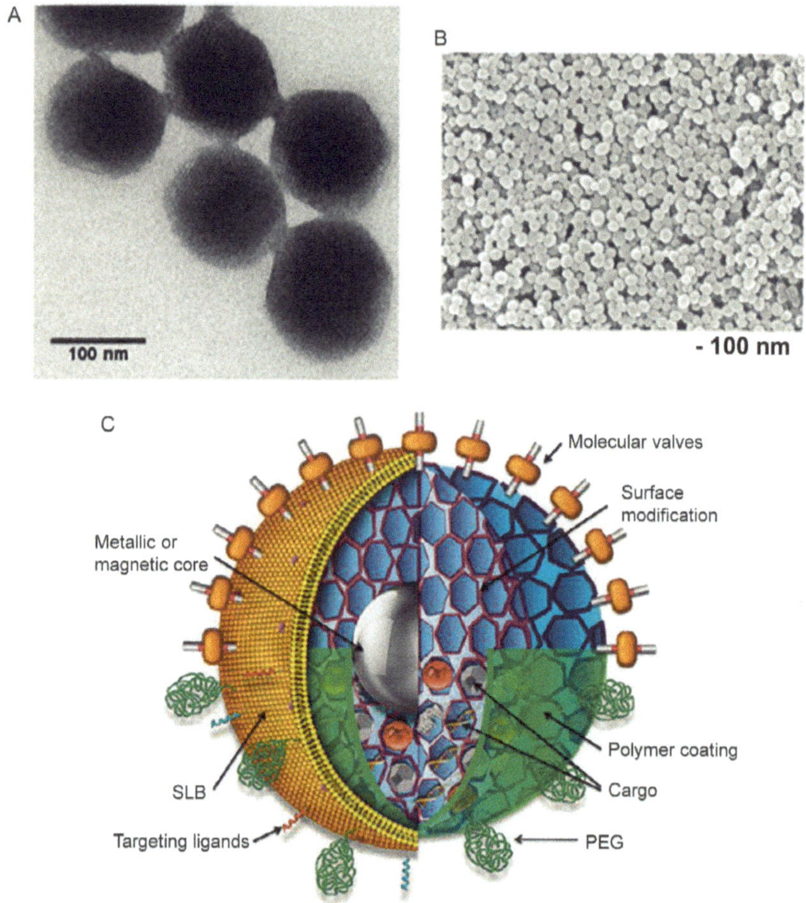

Fig. 4 Mesoporous silica nanoparticle structure. (A) Electron micrograph showing mesoporous silica nanoparticles (MSNs) of ~120 nm in size. Close examination reveals honeycomb-like pore structure. (B) MSNs in a batch are very uniform in size. (C) Schematic showing many of the elements and modifications that can be made to MSNs, including valves for the control of drug loading and metallic cores for magnetic field sensitivity [82].

We have recently shown *in vivo* efficacy of MSN-delivered siRNA in a mouse model of melanoma. Mice treated with siRNA against *TWIST* had smaller, less vascularized tumors, likely the result of inhibition of CCL2-driven angiogenesis within the growing tumors [71]. Moreover, similar data were also shown in ovarian cancer, where mice treated with siTWIST-MSN plus chemotherapy had significantly less tumor burden by 75% compared to control chemotherapy only treated mice [95]. Chapter "Synthesis and optical applications of periodic mesoporous organosilicas" by Inagaki further demonstrates *in vivo* efficacy of MSNs in a xenograft model of drug resistant ovarian cancer.

The next generation of MSNs may take advantage of their highly customizable nature. It is possible to combine PEI coating, targeting moieties, and/or the ability to carry drugs in their interior to create multifunctional particles, and much work remains to be done to realize the full potential of these particles.

8. THE POTENTIAL TO EXPLOIT: HYALURONIC ACID FOR CSC TARGETED THERAPY ON MSNs

8.1 The CSC Is an Attractive Therapeutic Target

Cluster of differentiation 44 (CD44) is a ubiquitously present glycoprotein on the surface of mammalian cells that plays a significant role in a number of biological functions [96–98]. Since the discovery that the receptor is overexpressed in a variety of solid tumors, such as pancreatic, breast and lung cancer, many studies have focused on methods for targeting CD44 in an attempt to improve drug delivery and discrimination between healthy and malignant tissue, while reducing residual toxicity and off-target accumulation [96–98]. Thus, CD44 receptor has great involvement in the different stages of tumor growth and metastasis, as well as methods currently used for targeting the receptor. Hyaluronic acid, the primary CD44 binding molecule, has proved a significant ally in developing nanocarriers that demonstrate preferential tumor accumulation and increased cell uptake [89,90]. This work sought to take advantage of hyaluronic acid's targeting ability and describes the possible advantages for each approach. The value of CD44 targeting can be easily appreciated from the number of different approaches that have reached clinical trials. On this note, CD44+ targeted delivery of HA-Coated mesoporous silica nanoparticles (MSNs) has been studied as a therapeutic approach to deliver oligonucleotides to tumors [57] (Fig. 5). We have overcome many difficulties [99] associated with delivery of small inhibitory RNA (siRNA) to tumors by MSN. Using MSN, we delivered Twist siRNA to breast and ovarian cancer xenografts and reduced tumor burden [71,95,100], opening up a new era of using siRNA to shut down EMT in a variety of cancers. In order to overcome cancer cell self-renewal and tumor-propagating potential, MSNs conjugated with CSC targeted hyaluronic acid show to inhibit EMT signaling, reduce tumor burden, and exhibit synergistic efficacy in combination with cisplatin [100], thus providing a useful therapeutic strategy for chemoresistant ovarian cancer.

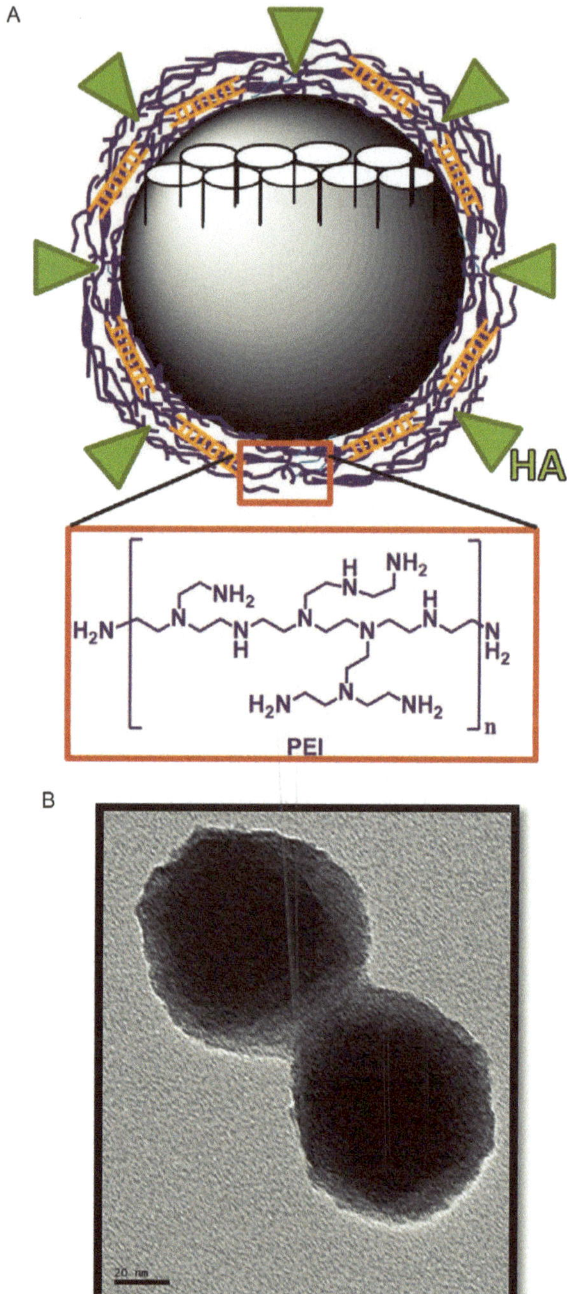

Fig. 5 (A) Schematic of an MSN-HA with pore structure (white). MSN-HAs used in these studies have a PEI coating (purple layer) which allow for a balance of charge interactions to chemically bind siRNA (orange) and hyaluronic acid (HA) (green). Monomer structure for PEI is shown below (red box). (B) Transmission electron micrograph of MSN-HAs. Particles are of uniform diameter, ~120 nm.

9. SUMMARY

The potential therapeutic value of siRNA-based nanoparticle therapies targeting Twist combined with cancer stem cell inhibitors like siSnail and let-7 has yet to be fully realized. If efficacious using *in vitro* and *in vivo* models of breast and ovarian cancer, these novel approaches may reach clinical trials within the next 5 years, which, if successful, could dramatically improve long-term outcomes for BC and OC patients and enable us to move closer to translating much needed therapeutic approaches into the clinic. Additionally, successful application of these types of approaches could pave the way for future RNA-based therapies against other targets of interest currently thought to be "undruggable."

REFERENCES

[1] A.J. Cortez, P. Tudrej, K.A. Kujawa, K.M. Lisowska, Advances in ovarian cancer therapy, Cancer Chemother. Pharmacol. 81 (2018) 17–38.
[2] N.M. Berezhnaya, Interaction between tumor and immune system: the role of tumor cell biology, Exp. Oncol. 32 (2010) 159–166.
[3] A.L. Jackson, E.L. Eisenhauer, T.J. Herzog, Emerging therapies: angiogenesis inhibitors for ovarian cancer, Expert Opin. Emerg. Drugs 20 (2015) 331–346.
[4] U. Solmaz, E. Mat, M. Levent Dereli, V. Turan, N. Peker, G. Tosun, A. Dogan, M. Adiyeke, A. Ozdemir, K. Gungorduk, M. Sanci, Y. Yildirim, Does neoadjuvant chemotherapy plus cytoreductive surgery improve survival rates in patients with advanced epithelial ovarian cancer compared with cytoreductive surgery alone? J. BUON 20 (2015) 580–587.
[5] R. Cornelison, D.C. Llaneza, C.N. Landen, Emerging therapeutics to overcome chemoresistance in epithelial ovarian cancer: a mini-review, Int. J. Mol. Sci. 18 (10) (2017) 54.
[6] J. Deng, L. Wang, H. Chen, J. Hao, J. Ni, L. Chang, W. Duan, P. Graham, Y. Li, Targeting epithelial-mesenchymal transition and cancer stem cells for chemoresistant ovarian cancer, Oncotarget 7 (2016) 55771–55788.
[7] C. DeSantis, J. Ma, L. Bryan, A. Jemal, Breast cancer statistics, 2013, CA Cancer J. Clin. 64 (2014) 52–62.
[8] R. Siegel, D. Naishadham, A. Jemal, Cancer statistics, 2012, CA Cancer J. Clin. 62 (2012) 10–29.
[9] A.D. Al Sayed, A.N. El Weshi, A.M. Tulbah, M.M. Rahal, A.A. Ezzat, Metaplastic carcinoma of the breast clinical presentation, treatment results and prognostic factors, Acta Oncol. 45 (2006) 188–195.
[10] A. Messere, E. Fabbri, M. Borgatti, R. Gambari, B. Di Blasio, C. Pedone, A. Romanelli, Antiproliferative activity of Pt(II) and Pd(II) phosphine complexes with thymine and thymidine, J. Inorg. Biochem. 101 (2007) 254–260.
[11] F. Vesuna, A. Lisok, B. Kimble, V. Raman, Twist modulates breast cancer stem cells by transcriptional regulation of CD24 expression, Neoplasia 11 (2009) 1318–1328.
[12] F. Vesuna, P. van Diest, J.H. Chen, V. Raman, Twist is a transcriptional repressor of E-cadherin gene expression in breast cancer, Biochem. Biophys. Res. Commun. 367 (2008) 235–241.

[13] B.G. Hollier, K. Evans, S.A. Mani, The epithelial-to-mesenchymal transition and cancer stem cells: a coalition against cancer therapies, J. Mammary Gland Biol. Neoplasia 14 (2009) 29–43.
[14] D. Kong, Y. Li, Z. Wang, F.H. Sarkar, Cancer stem cells and epithelial-to-mesenchymal transition (EMT)-phenotypic cells: are they cousins or twins? Cancers (Basel) 3 (2011) 716–729.
[15] N. Takebe, R.Q. Warren, S.P. Ivy, Breast cancer growth and metastasis: interplay between cancer stem cells, embryonic signaling pathways and epithelial-to-mesenchymal transition, Breast Cancer Res. 13 (2011) 211.
[16] S. Ansieau, J. Bastid, A. Doreau, A.P. Morel, B.P. Bouchet, C. Thomas, F. Fauvet, I. Puisieux, C. Doglioni, S. Piccinin, R. Maestro, T. Voeltzel, A. Selmi, S. Valsesia-Wittmann, C. Caron de Fromentel, A. Puisieux, Induction of EMT by twist proteins as a collateral effect of tumor-promoting inactivation of premature senescence, Cancer Cell 14 (2008) 79–89.
[17] O. Watanabe, H. Imamura, T. Shimizu, J. Kinoshita, T. Okabe, A. Hirano, K. Yoshimatsu, S. Konno, M. Aiba, K. Ogawa, Expression of twist and wnt in human breast cancer, Anticancer Res. 24 (2004) 3851–3856.
[18] A.E. Vernon, C. LaBonne, Tumor metastasis: a new twist on epithelial-mesenchymal transitions, Curr. Biol. 14 (2004) R719–R721.
[19] M.Y. Feng, K. Wang, H.T. Song, H.W. Yu, Y. Qin, Q.T. Shi, J.S. Geng, Metastasis-induction and apoptosis-protection by TWIST in gastric cancer cells, Clin. Exp. Metastasis 26 (2009) 1013–1023.
[20] R. Maestro, A.P. Dei Tos, Y. Hamamori, S. Krasnokutsky, V. Sartorelli, L. Kedes, C. Doglioni, D.H. Beach, G.J. Hannon, Twist is a potential oncogene that inhibits apoptosis, Genes Dev. 13 (1999) 2207–2217.
[21] J. Fu, L. Zhang, T. He, X. Xiao, X. Liu, L. Wang, L. Yang, M. Yang, T. Zhang, R. Chen, J. Xu, TWIST represses estrogen receptor-alpha expression by recruiting the NuRD protein complex in breast cancer cells, Int. J. Biol. Sci. 8 (2012) 522–532.
[22] F. Vesuna, A. Lisok, B. Kimble, J. Domek, Y. Kato, P. van der Groep, D. Artemov, J. Kowalski, H. Carraway, P. van Diest, V. Raman, Twist contributes to hormone resistance in breast cancer by downregulating estrogen receptor-alpha, Oncogene 31 (2011) 3223–3234.
[23] M.J. Fackler, M. McVeigh, E. Evron, E. Garrett, J. Mehrotra, K. Polyak, S. Sukumar, P. Argani, DNA methylation of RASSF1A, HIN-1, RAR-beta, Cyclin D2 and Twist in in situ and invasive lobular breast carcinoma, Int. J. Cancer 107 (2003) 970–975.
[24] D. Cakouros, S. Isenmann, L. Cooper, A. Zannettino, P. Anderson, C. Glackin, S. Gronthos, Twist-1 induces Ezh2 recruitment regulating histone methylation along the Ink4A/Arf locus in mesenchymal stem cells, Mol. Cell. Biol. 32 (2012) 1433–1441.
[25] T.A. Martin, A. Goyal, G. Watkins, W.G. Jiang, Expression of the transcription factors snail, slug, and twist and their clinical significance in human breast cancer, Ann. Surg. Oncol. 12 (2005) 488–496.
[26] J.G. van Nes, E.M. de Kruijf, H. Putter, D. Faratian, A. Munro, F. Campbell, V.T. Smit, G.J. Liefers, P.J. Kuppen, C.J. van de Velde, J.M. Bartlett, Co-expression of SNAIL and TWIST determines prognosis in estrogen receptor-positive early breast cancer patients, Breast Cancer Res. Treat. 133 (2012) 49–59.
[27] S. Li, S.E. Kendall, R. Raices, J. Finlay, M. Covarrubias, Z. Liu, G. Lowe, Y.H. Lin, Y.H. Teh, V. Leigh, S. Dhillon, S. Flanagan, K.S. Aboody, C.A. Glackin, TWIST1 associates with NF-kappaB subunit RELA via carboxyl-terminal WR domain to promote cell autonomous invasion through IL8 production, BMC Biol. 10 (2012) 73.
[28] D. Cakouros, R.M. Raices, S. Gronthos, C.A. Glackin, Twist-ing cell fate: mechanistic insights into the role of twist in lineage specification/differentiation and tumorigenesis, J. Cell. Biochem. 110 (2010) 1288–1298.

[29] Q.Q. Li, J.D. Xu, W.J. Wang, X.X. Cao, Q. Chen, F. Tang, Z.Q. Chen, X.P. Liu, Z.D. Xu, Twist1-mediated adriamycin-induced epithelial-mesenchymal transition relates to multidrug resistance and invasive potential in breast cancer cells, Clin. Cancer Res. 15 (2009) 2657–2665.
[30] A. Vichalkovski, E. Gresko, D. Hess, D.F. Restuccia, B.A. Hemmings, PKB/AKT phosphorylation of the transcription factor Twist-1 at Ser42 inhibits p53 activity in response to DNA damage, Oncogene 29 (2010) 3554–3565.
[31] M. Shiota, A. Yokomizo, M. Itsumi, T. Uchiumi, Y. Tada, Y. Song, E. Kashiwagi, D. Masubuchi, S. Naito, Twist1 and Y-box-binding protein-1 promote malignant potential in bladder cancer cells, BJU Int. 108 (2011) E142–E149.
[32] C.A. Glackin, E.J. Murray, S.S. Murray, Doxorubicin inhibits differentiation and enhances expression of the helix-loop-helix genes Id and mTwi in mouse osteoblastic cells, Biochem. Int. 28 (1992) 67–75.
[33] J. Dupont, A.M. Fernandez, C.A. Glackin, L. Helman, D. LeRoith, Insulin-like growth factor 1 (IGF-1)-induced twist expression is involved in the anti-apoptotic effects of the IGF-1 receptor, J. Biol. Chem. 276 (2001) 26699–26707.
[34] L. Oliveira-Ferrer, K. Rossler, V. Haustein, C. Schroder, D. Wicklein, D. Maltseva, N. Khaustova, T. Samatov, A. Tonevitsky, S. Mahner, F. Janicke, U. Schumacher, K. Milde-Langosch, c-FOS suppresses ovarian cancer progression by changing adhesion, Br. J. Cancer 110 (2014) 753–763.
[35] Q. Qin, Y. Xu, T. He, C. Qin, J. Xu, Normal and disease-related biological functions of Twist1 and underlying molecular mechanisms, Cell Res. 22 (2012) 90–106.
[36] J. Yang, S.A. Mani, J.L. Donaher, S. Ramaswamy, R.A. Itzykson, C. Come, P. Savagner, I. Gitelman, A. Richardson, R.A. Weinberg, Twist, a master regulator of morphogenesis, plays an essential role in tumor metastasis, Cell 117 (2004) 927–939.
[37] J. Yang, S.A. Mani, R.A. Weinberg, Exploring a new twist on tumor metastasis, Cancer Res. 66 (2006) 4549–4552.
[38] W.S. Wang, X.S. Yang, M. Xia, H.Y. Jiang, J.Q. Hou, Silencing of twist expression by RNA interference suppresses epithelial-mesenchymal transition, invasion, and metastasis of ovarian cancer, Asian Pac. J. Cancer Prev. 13 (2012) 4435–4439.
[39] C. Fan, A. Prat, J.S. Parker, Y. Liu, L.A. Carey, M.A. Troester, C.M. Perou, Building prognostic models for breast cancer patients using clinical variables and hundreds of gene expression signatures, BMC Med. Genet. 4 (2011) 3.
[40] J.Y. Lou, Z.L. Peng, Y. Zheng, H. Wang, B. He, H.J. Wang, Reversal of multi-drug resistance in ovarian cancer cell by RNA interference, Zhonghua Fu Chan Ke Za Zhi 41 (2006) 413–416.
[41] H. Yamaguchi, Y. Du, K. Nakai, M. Ding, S.S. Chang, J.L. Hsu, J. Yao, Y. Wei, L. Nie, S. Jiao, W.C. Chang, C.H. Chen, Y. Yu, G.N. Hortobagyi, M.C. Hung, EZH2 contributes to the response to PARP inhibitors through its PARP-mediated poly-ADP ribosylation in breast cancer, Oncogene 37 (2018) 208–217.
[42] S.M. Domchek, C. Aghajanian, R. Shapira-Frommer, R.K. Schmutzler, M.W. Audeh, M. Friedlander, J. Balmana, G. Mitchell, G. Fried, S.M. Stemmer, A. Hubert, O. Rosengarten, N. Loman, J.D. Robertson, H. Mann, B. Kaufman, Efficacy and safety of olaparib monotherapy in germline BRCA1/2 mutation carriers with advanced ovarian cancer and three or more lines of prior therapy, Gynecol. Oncol. 140 (2016) 199–203.
[43] J.F. Liu, U.A. Matulonis, What is the place of PARP inhibitors in ovarian cancer treatment? Curr. Oncol. Rep. 18 (2016) 29.

[44] M.R. Mirza, B.J. Monk, J. Herrstedt, A.M. Oza, S. Mahner, A. Redondo, M. Fabbro, J.A. Ledermann, D. Lorusso, I. Vergote, N.E. Ben-Baruch, C. Marth, R. Madry, R.D. Christensen, J.S. Berek, A. Dorum, A.V. Tinker, A. du Bois, A. Gonzalez-Martin, P. Follana, B. Benigno, P. Rosenberg, L. Gilbert, B.J. Rimel, J. Buscema, J.P. Balser, S. Agarwal, U.A. Matulonis, E.-O.N. Investigators, Niraparib maintenance therapy in platinum-sensitive, recurrent ovarian cancer, N. Engl. J. Med. 375 (2016) 2154–2164.
[45] J.A. Ledermann, PARP inhibitors in ovarian cancer, Ann. Oncol. 27 (Suppl. 1) (2016) i40–i44.
[46] M.W. Audeh, J. Carmichael, R.T. Penson, M. Friedlander, B. Powell, K.M. Bell-McGuinn, C. Scott, J.N. Weitzel, A. Oaknin, N. Loman, K. Lu, R.K. Schmutzler, U. Matulonis, M. Wickens, A. Tutt, Oral poly(ADP-ribose) polymerase inhibitor olaparib in patients with BRCA1 or BRCA2 mutations and recurrent ovarian cancer: a proof-of-concept trial, Lancet 376 (2010) 245–251.
[47] K. Masui, B. Gini, J. Wykosky, C. Zanca, P.S. Mischel, F.B. Furnari, W.K. Cavenee, A tale of two approaches: complementary mechanisms of cytotoxic and targeted therapy resistance may inform next-generation cancer treatments, Carcinogenesis 34 (2013) 725–738.
[48] L.N. Abdullah, E.K. Chow, Mechanisms of chemoresistance in cancer stem cells, Clin. Transl. Med. 2 (2013) 3.
[49] T.R. McPhee, P.C. McDonald, A. Oloumi, S. Dedhar, Integrin-linked kinase regulates E-cadherin expression through PARP-1, Dev. Dyn. 237 (2008) 2737–2747.
[50] G. Mariano, M.R. Ricciardi, D. Trisciuoglio, M. Zampieri, F. Ciccarone, T. Guastafierro, R. Calabrese, E. Valentini, A. Tafuri, D. Del Bufalo, P. Caiafa, A. Reale, PARP inhibitor ABT-888 affects response of MDA-MB-231 cells to doxorubicin treatment, targeting Snail expression, Oncotarget 6 (2015) 15008–15021.
[51] M.I. Rodriguez, A. Gonzalez-Flores, F. Dantzer, J. Collard, A.G. de Herreros, F.J. Oliver, Poly(ADP-ribose)-dependent regulation of Snail1 protein stability, Oncogene 30 (2011) 4365–4372.
[52] C.A. Doege, K. Inoue, T. Yamashita, D.B. Rhee, S. Travis, R. Fujita, P. Guarnieri, G. Bhagat, W.B. Vanti, A. Shih, R.L. Levine, S. Nik, E.I. Chen, A. Abeliovich, Early-stage epigenetic modification during somatic cell reprogramming by Parp1 and Tet2, Nature 488 (2012) 652–655.
[53] S.H. Chiou, B.H. Jiang, Y.L. Yu, S.J. Chou, P.H. Tsai, W.C. Chang, L.K. Chen, L.H. Chen, Y. Chien, G.Y. Chiou, Poly(ADP-ribose) polymerase 1 regulates nuclear reprogramming and promotes iPSC generation without c-Myc, J. Exp. Med. 210 (2013) 85–98.
[54] Y.S. Lai, C.W. Chang, K.M. Pawlik, D. Zhou, M.B. Renfrow, T.M. Townes, SRY (sex determining region Y)-box2 (Sox2)/poly ADP-ribose polymerase 1 (Parp1) complexes regulate pluripotency, Proc. Natl. Acad. Sci. U. S. A. 109 (2012) 3772–3777.
[55] B.H. Jiang, W.L. Tseng, H.Y. Li, M.L. Wang, Y.L. Chang, Y.J. Sung, S.H. Chiou, Poly (ADP-ribose) polymerase 1: cellular pluripotency, reprogramming, and tumorogenesis, Int. J. Mol. Sci. 16 (2015) 15531–15545.
[56] M.E. Wielgos, R. Rajbhandari, T.S. Cooper, S. Wei, S.E. Nozell, E.S. Yang, Let-7 status is crucial for PARP1 expression in HER2-overexpressing breast tumors, Mol. Cancer Res. 15 (2017) 340–347.
[57] H. Meng, W.X. Mai, H. Zhang, M. Xue, T. Xia, S. Lin, X. Wang, Y. Zhao, Z. Ji, J.I. Zink, A.E. Nel, Codelivery of an optimal drug/siRNA combination using mesoporous silica nanoparticles to overcome drug resistance in breast cancer in vitro and in vivo, ACS Nano 7 (2013) 994–1005.
[58] A. Fire, S. Xu, M.K. Montgomery, S.A. Kostas, S.E. Driver, C.C. Mello, Potent and specific genetic interference by double-stranded RNA in Caenorhabditis elegans, Nature 391 (1998) 806–811.

[59] E.J. Sontheimer, Assembly and function of RNA silencing complexes, Nat. Rev. Mol. Cell Biol. 6 (2005) 127–138.
[60] S.M. Elbashir, J. Harborth, W. Lendeckel, A. Yalcin, K. Weber, T. Tuschl, Duplexes of 21-nucleotide RNAs mediate RNA interference in cultured mammalian cells, Nature 411 (2001) 494–498.
[61] E. Bernstein, A.A. Caudy, S.M. Hammond, G.J. Hannon, Role for a bidentate ribonuclease in the initiation step of RNA interference, Nature 409 (2001) 363–366.
[62] M. Tijsterman, R.H. Plasterk, Dicers at RISC; the mechanism of RNAi, Cell 117 (2004) 1–3.
[63] J. Liu, M.A. Carmell, F.V. Rivas, C.G. Marsden, J.M. Thomson, J.J. Song, S.M. Hammond, L. Joshua-Tor, G.J. Hannon, Argonaute2 is the catalytic engine of mammalian RNAi, Science 305 (2004) 1437–1441.
[64] Z.W. Liang, B.F. Guo, Y. Li, X.J. Li, X. Li, L.J. Zhao, L.F. Gao, H. Yu, X.J. Zhao, L. Zhang, B.X. Yang, Plasmid-based Stat3 siRNA delivered by hydroxyapatite nanoparticles suppresses mouse prostate tumour growth in vivo, Asian J. Androl. 13 (2011) 481–486.
[65] M.E. Davis, J.E. Zuckerman, C.H. Choi, D. Seligson, A. Tolcher, C.A. Alabi, Y. Yen, J.D. Heidel, A. Ribas, Evidence of RNAi in humans from systemically administered siRNA via targeted nanoparticles, Nature 464 (2010) 1067–1070.
[66] A. Forsbach, J.G. Nemorin, C. Montino, C. Muller, U. Samulowitz, A.P. Vicari, M. Jurk, G.K. Mutwiri, A.M. Krieg, G.B. Lipford, J. Vollmer, Identification of RNA sequence motifs stimulating sequence-specific TLR8-dependent immune responses, J. Immunol. 180 (2008) 3729–3738.
[67] M. Robbins, A. Judge, I. MacLachlan, siRNA and innate immunity, Oligonucleotides 19 (2009) 89–102.
[68] M.A. Behlke, Chemical modification of siRNAs for in vivo use, Oligonucleotides 18 (2008) 305–319.
[69] W.F. Lima, H. Wu, J.G. Nichols, H. Sun, H.M. Murray, S.T. Crooke, Binding and cleavage specificities of human argonaute2, J. Biol. Chem. 284 (2009) 26017–26028.
[70] H.J. Haringsma, J.J. Li, F. Soriano, D.M. Kenski, W.M. Flanagan, A.T. Willingham, mRNA knockdown by single strand RNA is improved by chemical modifications, Nucleic Acids Res. 40 (2012) 4125–4136.
[71] J. Finlay, C.M. Roberts, J. Dong, J.I. Zink, F. Tamanoi, C.A. Glackin, Mesoporous silica nanoparticle delivery of chemically modified siRNA against TWIST1 leads to reduced tumor burden, Nanomedicine 11 (2015) 1657–1666.
[72] C. Foged, siRNA delivery with lipid-based systems: promises and pitfalls, Curr. Top. Med. Chem. 12 (2012) 97–107.
[73] M. Gooding, L.P. Browne, F.M. Quinteiro, D.L. Selwood, siRNA delivery: from lipids to cell-penetrating peptides and their mimics, Chem. Biol. Drug Des. 80 (2012) 787–809.
[74] S.K. Lee, A. Siefert, J. Beloor, T.M. Fahmy, P. Kumar, Cell-specific siRNA delivery by peptides and antibodies, Methods Enzymol. 502 (2012) 91–122.
[75] J. Wu, J. Zhou, F. Qu, P. Bao, Y. Zhang, L. Peng, Polycationic dendrimers interact with RNA molecules: polyamine dendrimers inhibit the catalytic activity of Candida ribozymes, Chem. Commun. (Camb.) (3) (2005) 313–315. (Epub 2004 Dec 9).
[76] X.X. Liu, P. Rocchi, F.Q. Qu, S.Q. Zheng, Z.C. Liang, M. Gleave, J. Iovanna, L. Peng, PAMAM dendrimers mediate siRNA delivery to target Hsp27 and produce potent antiproliferative effects on prostate cancer cells, ChemMedChem 4 (2009) 1302–1310.
[77] J. Zhou, C.P. Neff, X. Liu, J. Zhang, H. Li, D.D. Smith, P. Swiderski, T. Aboellail, Y. Huang, Q. Du, Z. Liang, L. Peng, R. Akkina, J.J. Rossi, Systemic administration of combinatorial dsiRNAs via nanoparticles efficiently suppresses HIV-1 infection in humanized mice, Mol. Ther. 19 (2011) 2228–2238.

[78] X.C. Shen, J. Zhou, X. Liu, J. Wu, F. Qu, Z.L. Zhang, D.W. Pang, G. Quelever, C.C. Zhang, L. Peng, Importance of size-to-charge ratio in construction of stable and uniform nanoscale RNA/dendrimer complexes, Org. Biomol. Chem. 5 (2007) 3674–3681.
[79] G.M. Pavan, P. Posocco, A. Tagliabue, M. Maly, A. Malek, A. Danani, E. Ragg, C.V. Catapano, S. Pricl, PAMAM dendrimers for siRNA delivery: computational and experimental insights, Chemistry 16 (2010) 7781–7795.
[80] J. Wu, W. Huang, Z. He, Dendrimers as carriers for siRNA delivery and gene silencing: a review, Sci. World J. 2013 (2013) 630654.
[81] R. Han, Y. Sun, C. Kang, H. Sun, W. Wei, Amphiphilic dendritic nanomicelle-mediated co-delivery of 5-fluorouracil and doxorubicin for enhanced therapeutic efficacy, J. Drug Target. 25 (2017) 140–148.
[82] J. Finlay, C.M. Roberts, G. Lowe, J. Loeza, J.J. Rossi, C.A. Glackin, RNA-based TWIST1 inhibition via dendrimer complex to reduce breast cancer cell metastasis, Biomed. Res. Int. 2015 (2015) 382745.
[83] P.C. Naha, M. Davoren, F.M. Lyng, H.J. Byrne, Reactive oxygen species (ROS) induced cytokine production and cytotoxicity of PAMAM dendrimers in J774A.1 cells, Toxicol. Appl. Pharmacol. 246 (2010) 91–99.
[84] W. Wang, W. Xiong, J. Wan, X. Sun, H. Xu, X. Yang, The decrease of PAMAM dendrimer-induced cytotoxicity by PEGylation via attenuation of oxidative stress, Nanotechnology 20 (2009) 105103.
[85] G. Nyitrai, L. Heja, I. Jablonkai, I. Pal, J. Visy, J. Kardos, Polyamidoamine dendrimer impairs mitochondrial oxidation in brain tissue, J. Nanobiotechnol. 11 (2013) 9.
[86] T.C. Heiden, E. Dengler, W.J. Kao, W. Heideman, R.E. Peterson, Developmental toxicity of low generation PAMAM dendrimers in zebrafish, Toxicol. Appl. Pharmacol. 225 (2007) 70–79.
[87] T. Yu, X. Liu, A.L. Bolcato-Bellemin, Y. Wang, C. Liu, P. Erbacher, F. Qu, P. Rocchi, J.P. Behr, L. Peng, An amphiphilic dendrimer for effective delivery of small interfering RNA and gene silencing in vitro and in vivo, Angew. Chem. Int. Ed. Eng. 51 (2012) 8478–8484.
[88] X. Liu, J. Zhou, T. Yu, C. Chen, Q. Cheng, K. Sengupta, Y. Huang, H. Li, C. Liu, Y. Wang, P. Posocco, M. Wang, Q. Cui, S. Giorgio, M. Fermeglia, F. Qu, S. Pricl, Y. Shi, Z. Liang, P. Rocchi, J.J. Rossi, L. Peng, Adaptive amphiphilic dendrimer-based nanoassemblies as robust and versatile siRNA delivery systems, Angew. Chem. Int. Ed. Eng. 53 (2014) 11822–11827.
[89] J. Lu, M. Liong, J.I. Zink, F. Tamanoi, Mesoporous silica nanoparticles as a delivery system for hydrophobic anticancer drugs, Small 3 (2007) 1341–1346.
[90] C. Hom, J. Lu, M. Liong, H. Luo, Z. Li, J.I. Zink, F. Tamanoi, Mesoporous silica nanoparticles facilitate delivery of siRNA to shutdown signaling pathways in mammalian cells, Small 6 (2010) 1185–1190.
[91] H. Kobayashi, R. Watanabe, P.L. Choyke, Improving conventional enhanced permeability and retention (EPR) effects; what is the appropriate target? Theranostics 4 (2013) 81–89.
[92] K. Greish, Enhanced permeability and retention (EPR) effect for anticancer nanomedicine drug targeting, Methods Mol. Biol. 624 (2010) 25–37.
[93] Y. Yamazaki, M. Nango, M. Matsuura, Y. Hasegawa, M. Hasegawa, N. Oku, Polycation liposomes, a novel nonviral gene transfer system, constructed from cetylated polyethylenimine, Gene Ther. 7 (2000) 1148–1155.
[94] T. Xia, M. Kovochich, M. Liong, H. Meng, S. Kabehie, S. George, J.I. Zink, A.E. Nel, Polyethyleneimine coating enhances the cellular uptake of mesoporous silica nanoparticles and allows safe delivery of siRNA and DNA constructs, ACS Nano 3 (2009) 3273–3286.

[95] C.M. Roberts, S.A. Shahin, W. Wen, J.B. Finlay, J. Dong, R. Wang, T.H. Dellinger, J.I. Zink, F. Tamanoi, C.A. Glackin, Nanoparticle delivery of siRNA against TWIST to reduce drug resistance and tumor growth in ovarian cancer models, Nanomedicine 13 (2017) 965–976.

[96] M. Al-Hajj, M.S. Wicha, A. Benito-Hernandez, S.J. Morrison, M.F. Clarke, Prospective identification of tumorigenic breast cancer cells, Proc. Natl. Acad. Sci. U. S. A. 100 (2003) 3983–3988.

[97] C. Li, D.G. Heidt, P. Dalerba, C.F. Burant, L. Zhang, V. Adsay, M. Wicha, M.F. Clarke, D.M. Simeone, Identification of pancreatic cancer stem cells, Cancer Res. 67 (2007) 1030–1037.

[98] G. Yin, R. Chen, A.B. Alvero, H.H. Fu, J. Holmberg, C. Glackin, T. Rutherford, G. Mor, TWISTing stemness, inflammation and proliferation of epithelial ovarian cancer cells through MIR199A2/214, Oncogene 29 (2010) 3545–3553.

[99] P. Guo, O. Coban, N.M. Snead, J. Trebley, S. Hoeprich, S. Guo, Y. Shu, Engineering RNA for targeted siRNA delivery and medical application, Adv. Drug Deliv. Rev. 62 (2010) 650–666.

[100] S.A. Shahin, R. Wang, S.I. Simargi, A. Contreras, L. Parra Echavarria, L. Qu, W. Wen, T. Dellinger, J. Unternaehrer, F. Tamanoi, J.I. Zink, C.A. Glackin, Hyaluronic acid conjugated nanoparticle delivery of siRNA against TWIST reduces tumor burden and enhances sensitivity to cisplatin in ovarian cancer, Nanomedicine 14 (2018) 1381–1394.

CHAPTER SIX

Anticancer Drug Delivery Capability of Biodegradable PMO in the Chicken Egg Tumor Model

Kotaro Matsumoto*, Tan Le Hoang Doan*,†,‡, Ngoc Xuan Dat Mai*,†, Keigo Nakai*, Aoi Komatsu*, Fuyuhiko Tamanoi*,§,1

*Institute for Integrated Cell-Material Sciences, Institute for Advanced Study, Kyoto University, Kyoto, Japan
†Center for Innovative Materials and Architectures (INOMAR), University of Science, Vietnam National University—Ho Chi Minh City, Ho Chi Minh City, Vietnam
‡Faculty of Chemistry, University of Science, Vietnam National University—Ho Chi Minh City, Ho Chi Minh City, Vietnam
§Department of Microbiology, Immunology and Molecular Genetics, University of California, Los Angeles, CA, United States
[1]Corresponding author: e-mail address: fuyut@microbio.ucla.edu

Contents

Abstract

Biodegradable PMO (Periodic Mesoporous Organosilica) has emerged as a promising type of mesoporous silica based nanoparticles for clinical translation. We as well as others have synthesized and characterized three types of nanoparticles containing disulfide, tetrasulfide or peptide-like bonds within their framework. Biodegradable PMO has high drug loading efficiency and can be taken up into cancer cells. Our experiments utilizing the chicken egg tumor model uncovered excellent capability to accumulate and deliver anticancer drugs to the tumor. These results will be described and their significance will be discussed. The chicken egg tumor model provides a convenient and versatile model for characterizing nanoparticles.

The Enzymes, Volume 44
ISSN 1874-6047
https://doi.org/10.1016/bs.enz.2018.09.003
© 2018 Elsevier Inc.
All rights reserved.

1. INTRODUCTION

Mesoporous silica nanoparticles (MSNs) have been shown to provide a promising vehicle to deliver anticancer agents [1–4]. MSNs are biocompatible and safe, as examined by a number of mouse studies [5,6]. However, when contemplating clinical translation, it is critical that maximum effort is placed on guaranteeing safety of the material. While MSNs are degradable, it takes time to degrade and the rate of degradation may differ among different environment in the body. It is desirable that the nanoparticles completely disappear after accomplishing their mission to deliver anticancer drugs to the tumor. Thus, it has been a holy grail of the MSN research to tune biodegradability of the nanoparticles.

We as well as others have recently characterized a new type of MSN termed "Biodegradable PMO" that contains biodegradable bonds within the framework of the nanoparticles [7–10]. The biodegradable bonds employed include disulfide, tetrasulfide or peptide-like bonds that are expected to be degraded within the body. Studies so far suggest that the biodegradable PMO preserves all the capabilities of MSN including tumor accumulation and drug delivery. At least in in vitro studies, degradation of biodegradable PMO has been demonstrated. Thus, it is expected that complete elimination of the nanomaterial from animal body can be envisioned. Biodegradable silica based nanoparticles are discussed by Croissant and Brinker in the previous volume (Volume 43).

We have recently reported excellent tumor accumulation and tumor growth inhibition capability of biodegradable PMO nanoparticles by using the chicken egg tumor model [11]. In this chapter, we first describe these results and then discuss implication of the study for future potential clinical translation. We will also discuss a variety of applications of the chicken egg tumor model.

Animal models such as mouse tumor models have been widely used to examine the ability of nanoparticles to deliver anticancer drugs and inhibit tumor growth. The chicken egg tumor model represents an alternative animal model that uses fertilized eggs [12–18]. In this model, fertilized eggs are incubated for 10 days at which time cancer cells are transplanted onto the chorioallantoic membrane (CAM) that surrounds chicken embryo. The CAM membrane supports nutrient delivery. In addition, calcium is degraded from the egg shell and provided to the embryo. The CAM membrane provides a nutrient rich environment for tumor growth after transplanting cancer cells. This system provides a convenient and versatile animal model to study effect of nanoparticle-medicated anticancer drug delivery.

2. BIODEGRADABLE PMO NANOPARTICLES AS A PROMISING TYPE OF MESOPOROUS SILICA BASED NANOMATERIAL THAT IS ATTRACTIVE FOR CLINICAL TRANSLATION

Fig. 1 summarizes three types of biodegradable bonds that have been characterized [7–10]. They incorporate disulfide and tetrasulfide bonds that are degraded by reducing conditions as well as a peptide-like bond that can be degraded by proteases inside the body. As shown in Fig. 1B, biodegradable PMO having an oxamide bond can be degraded by the treatment with trypsin for 48 h [10]. It is expected that these small degradation products will be released into the urine and leave the body.

The method for the synthesis of biodegradable PMO is briefly described in Fig. 2. Panel A shows a typical synthesis of mesoporous silica nanoparticles that uses tetraalkoxysilane (TEOS) as a precursor. Condensation of TEOS occurs in alkaline conditions resulting in the formation of Si–O–Si bonds in the framework. Addition of surfactants such as CTAB and subsequent removal of the surfactant produces thousands of pores. For the synthesis

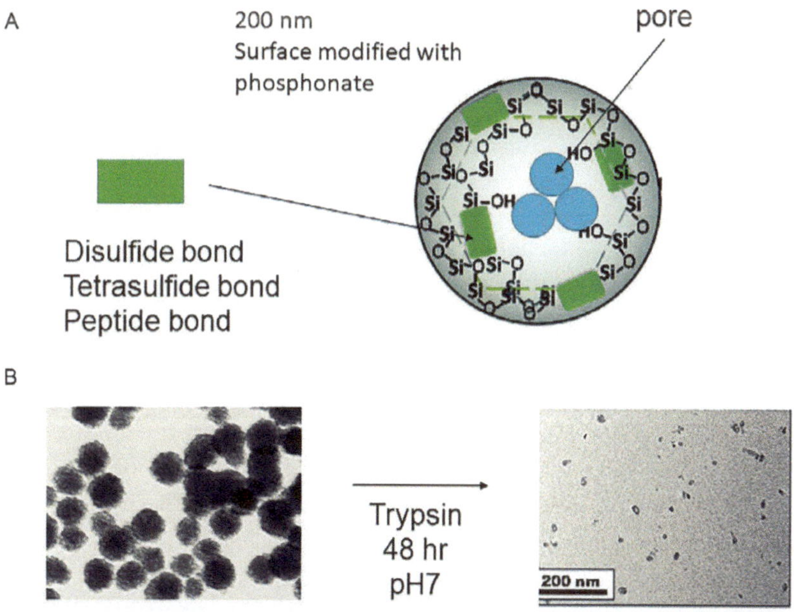

Fig. 1 Biodegradable PMO nanoparticles. (A) Graphic representation of biodegradable PMO nanoparticles that contain disulfide, tetrasulfide or peptide-like bonds. (B) Degradation of biodegradable PMO containing peptide-like bonds by the incubation with trypsin was confirmed by TEM.

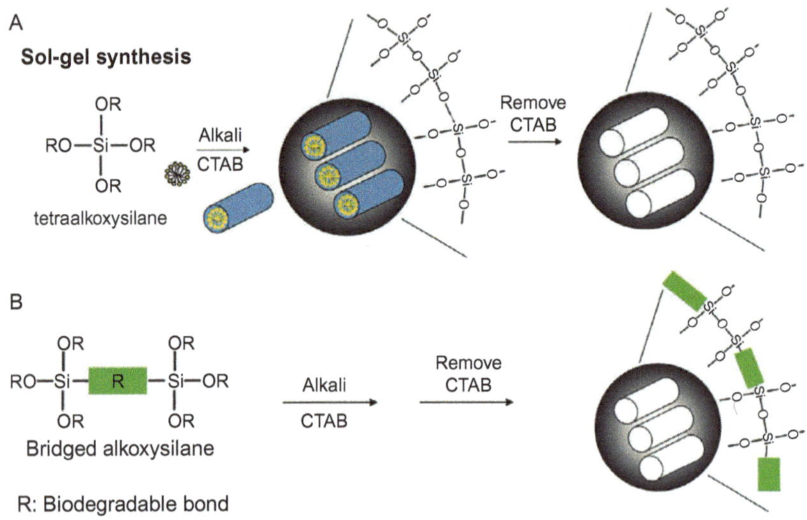

Fig. 2 Synthesis of biodegradable PMO. (A) Synthesis of mesoporous silica nanoparticles using sol-gel synthesis. Tetraalkoxysilane is condensed in alkali condition in the presence of surfactants such as CTAB. (B) Synthesis of biodegradable PMO using bridged alkoxysilane. Note that a variety of biodegradable bonds can be incorporated into R.

of biodegradable PMO, we use bridged alkoxysilane instead of TEOS. A biodegradable bond can be inserted between the two Si atoms. These bonds are retained during nanoparticle synthesis and are incorporated into the framework. It was found during the synthesis of a disulfide containing nanoparticles that it is necessary to mix the bridged disulfide alkoxysilane with another bridged precursor, ethylene-containing alkoxysilane. In other cases, ethane-containing alkoxysilane is used. This appears to be due to the necessity to strengthen the integrity of nanoparticles. Various ratios of the two bridged alkoxysilanes have been tested to gain an optimum ratio. The nanoparticles are also surface modified with phosphonate.

Anticancer drugs can be loaded onto biodegradable PMOs. With doxorubicin, using a low pH condition such as 5.5 increases the amount of the drug that can be loaded [7]. This results in the preparation of drug-loaded nanoparticle where the drug represents from 20% to 50% of total weight. The enhanced drug loading is believed to be due to interactions of the drug with organic components of the nanoparticle surface.

Similar to mesoporous silica nanoparticles, biodegradable PMOs are taken up into cancer cells and accumulate close to the nucleus, presumably reflecting their localization in lysosomes [7,10]. Delivery of anticancer drugs

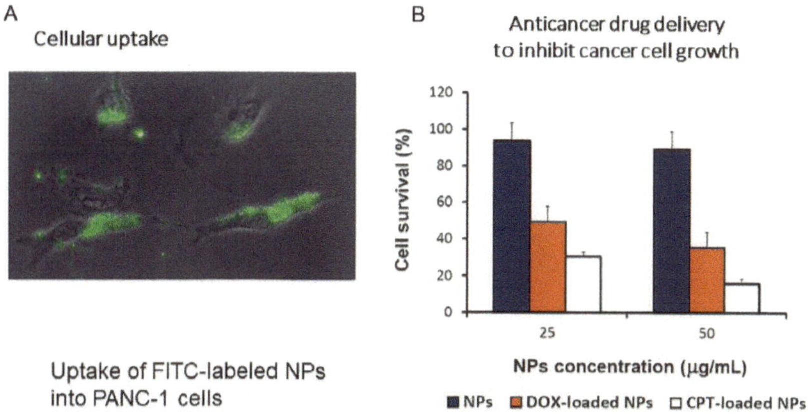

Fig. 3 In vitro study of drug-loaded biodegradable PMO. (A) Uptake of FITC-labeled biodegradable PMO containing disulfide bond into pancreatic cancer cells PANC-1. (B) Delivery of doxorubicin or camptothecin by biodegradable PMO results in decreased survival of PANC-1 cells. *Adapted from J.G. Croissant, Y. Fatieiev, K. Julfakyan, J. Lu, A.-H. Emwas, D.H. Anjum, H. Omar, F. Tamanoi, J.I. Zink, N.M. Khashab, Biodegradable oxamide-phenylene-based mesoporous organosilica nanoparticles with unprecedented drug payloads for delivery in cells, Chemistry 22 (42) (2016) 14806–14811.*

by loading doxorubicin or camptothecin in the nanoparticles results in killing of pancreatic cancer PANC-1 cells (Fig. 3). These results show that, like mesoporous silica nanoparticles, biodegradable PMO provides an effective delivery vehicle for anticancer drugs.

3. EVALUATION OF BIODEGRADABLE PMO BY THE USE OF CHICKEN EGG TUMOR MODEL UNCOVERED EXCELLENT CAPABILITY OF THESE NANOPARTICLES

To start evaluating the capacity of biodegradable PMO to deliver anticancer drugs to the tumor in an animal setting, we employed the chicken egg CAM assay as a simple and versatile model [11]. This assay uses fertilized eggs which are incubated for 10 days at 37 °C and 65% humidity (Fig. 4). At 10 days during fertilization, chick embryo develops and is surrounded by a nutrient rich membrane called chorioallantoic membrane. A window is made on the egg shell and human cancer cells are transplanted onto the CAM membrane. Surprisingly, tumor is formed in 3 days after the transplantation (Fig. 4). The tumor formed retains characteristics of tumor. A variety of human cancer cell lines can be used to form a tumor. We have used ovarian cancer, lung cancer, glioblastoma cells with success. H&E staining results

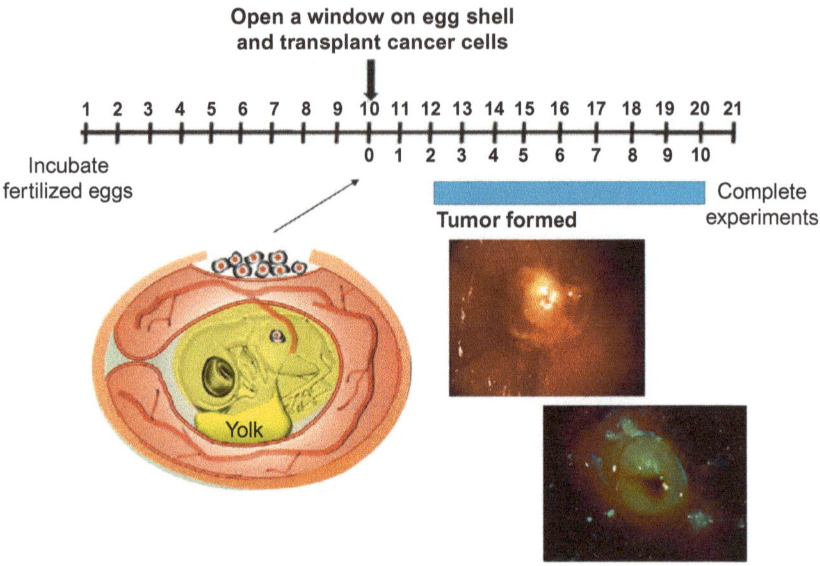

Fig. 4 Chicken egg tumor model. A general protocol for the chicken egg tumor model experiments. Fertilized eggs are incubated for 21 days during which time all the experiments will be completed. At day 10, a window is made on the egg shell and human cancer cells (or tumor sample) are placed on top of the CAM membrane. Tumor is formed 3 or 4 days after the transplantation. Examples of the tumor formed using ovarian cancer cells (upper) or melanoma cells (lower) are shown.

of tumor section made by transplanting human ovarian cells OVCAR8 or human lung cancer cells A549 are shown in Fig. 5. Different staining patterns representing features of each cancer are seen. Other staining such as trichrome staining identifies the presence of extracellular matrix (ECM). In addition, staining with anti-Vimentin demonstrates the presence of stromal cells. In addition, these tumors are highly vascularized. Thus, tumor characteristics can be identified in the CAM tumor.

The amount of nanoparticles injected intravenously significantly influences biodistribution and tumor accumulation. Therefore, we have first injected intravenously varying amounts (0.1–0.5 mg) of biodegradable PMOs labeled with fluorescent dye and examined tumor accumulation. It appears that injecting 0.2 mg results in significant accumulation of nanoparticles in the tumor. The tumor accumulation is seen 1 day after the injection and continues at least 3 days. Increasing the amount to 0.5 mg results in accumulation to liver and spleen in addition to tumor.

We next loaded biodegradable PMO with doxorubicin and injected intravenously into the chicken egg. As shown in Fig. 6, complete elimination

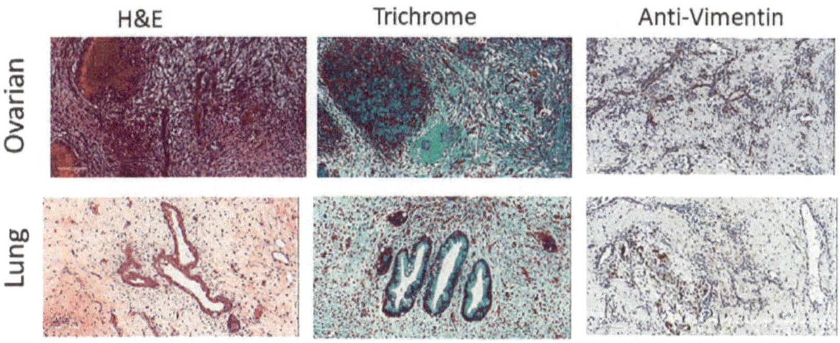

Fig. 5 Staining of the tumor section formed on the CAM membrane. Tumor formed on the CAM membrane by transplanting ovarian cancer cells (OVCAR8) or lung cancer cells (A549) is cut out and thin sections were prepared. They were stained by H&E staining, trichrome staining as well as by using anti-Vimentin antibody. *Adapted from V. Binh, S. Shahin, J. Croissant, Y. Fatieiev, K. Matsumoto, T. Le-Hoang Doan, T. Yik, S. Simargi, A. Conteras, L. Ratliff, C. Mauriello Jimenez, L. Raehm, N. Khashab, J.-O. Durand, C. Glackin, F. Tamanoi, Chick chorioallantoic membrane assay as an in vivo model to study the effect of nanoparticle-based anticancer drugs in ovarian cancer, Sci. Rep. 8 (2018).*

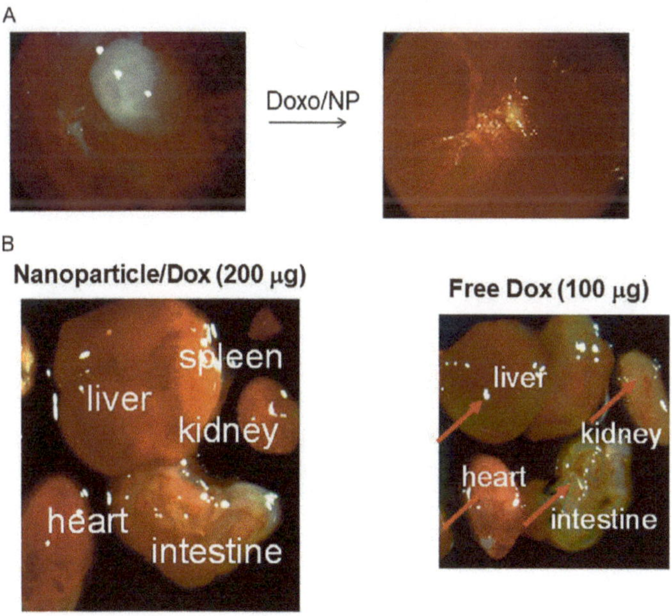

Fig. 6 Tumor elimination be injecting doxorubicin-loaded biodegradable PMO. (A) Tumor was formed on the CAM membrane by transplanting OVCAR8 cells. Doxorubicin-loaded biodegradable PMO was intravenously injected. Three days after the administration, elimination of the tumor was observed. (B) Lack of organ damage by the injection of doxorubicin-loaded biodegradable PMO (the amount of doxorubicin injected is 200 μg) (left panel). Tumor as well as various organs from the embryo were cut out and their appearance was examined. The right panel shows the same experiment when free doxorubicin (100 μg) was injected. In this case, severe organ damage was observed. *Adapted from V. Binh, S. Shahin, J. Croissant, Y. Fatieiev, K. Matsumoto, T. Le-Hoang Doan, T. Yik, S. Simargi, A. Conteras, L. Ratliff, C. Mauriello Jimenez, L. Raehm, N. Khashab, J.-O. Durand, C. Glackin, F. Tamanoi, Chick chorioallantoic membrane assay as an in vivo model to study the effect of nanoparticle-based anticancer drugs in ovarian cancer, Sci. Rep. 8 (2018).*

of tumor in 3–4 days was observed (Fig. 6A). The tumor elimination presumably reflects the ability of doxorubicin to induce apoptosis rather than causing cytostatic effect.

The CAM assay enables examination of internal organs of the embryo. Thus, organs such as liver, kidney, spleen, intestine, lung, stomach and heart were cut out and examined for their shape and color after injection of doxorubicin loaded biodegradable PMO nanoparticles. An important result was obtained when the results were compared with those obtained after injection of free doxorubicin. This point is clearly seen in Fig. 7A and B. Injection of free doxorubicin caused severe organ damage as observed

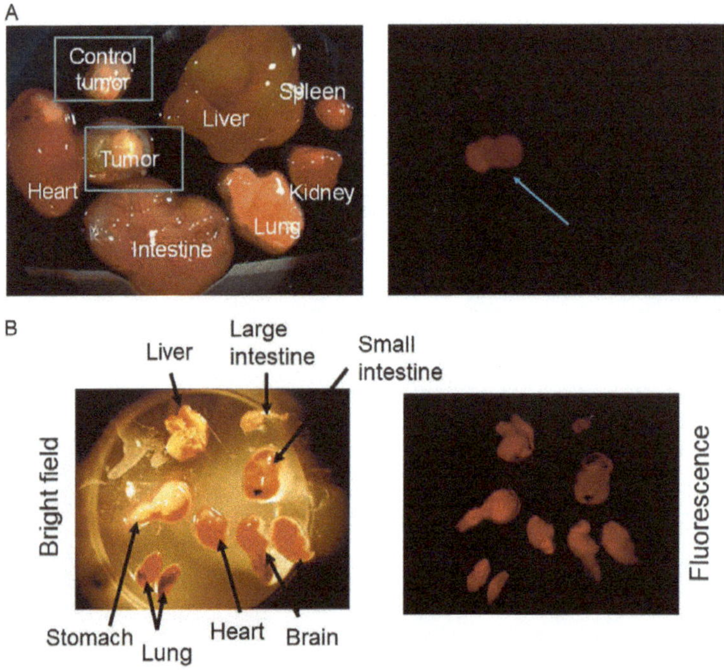

Fig. 7 Preferential delivery of doxorubicin to the tumor by nanoparticles. Ovarian tumor was formed on the CAM membrane. (A) Doxorubicin-loaded PMO nanoparticle was injected intravenously and the red fluorescence of doxorubicin was examined 3 days after the injection by cutting out the tumor as well as various organs. (B) Free doxorubicin was injected into the egg with ovarian tumor and distribution of doxorubicin was examined by following red fluorescence of the drug. *Adapted from V. Binh, S. Shahin, J. Croissant, Y. Fatieiev, K. Matsumoto, T. Le-Hoang Doan, T. Yik, S. Simargi, A. Conteras, L. Ratliff, C. Mauriello Jimenez, L. Raehm, N. Khashab, J.-O. Durand, C. Glackin, F. Tamanoi, Chick chorioallantoic membrane assay as an in vivo model to study the effect of nanoparticle-based anticancer drugs in ovarian cancer, Sci. Rep. 8 (2018).*

by the injection of 100 μg of the drug. In fact, 50 μg is the maximum amount that can be injected without causing organ damages. In contrast to this, at least 200 μg of doxorubicin could be injected when loaded onto biodegradable PMOs. The lack of organ damage when injecting a large amount of doxorubicin using nanoparticles was found to be due to preferential delivery of the drug to the tumor. As shown in Fig. 7A, red fluorescence of doxorubicin was detected in the tumor 1 or 2 days after injection of doxorubicin-loaded biodegradable PMO nanoparticles. These results contrast significantly with the results obtained after injecting free doxorubicin. In this case, red fluorescence of doxorubicin appeared in all organs examined (Fig. 7B right).

Preferential delivery of doxorubicin to the tumor appears to be due to the ability of nanoparticles to accumulate in the tumor. Presumably, circulation of nanoparticles in the blood stream allows nanoparticles to accumulate in the tumor and deliver doxorubicin specifically. In fact, when we intravenously injected FITC-labeled biodegradable PMO into the chicken egg with tumor, we observed preferential accumulation of green fluorescence in the tumor [11]. The nanoparticles used here do not have any targeting moiety on their surface. Thus, it appears that nanoparticles leak out of the blood stream at the tumor site by taking advantage of the leaky wall of the blood vessel (EPR effect) [19,20].

4. FUTURE PROSPECTS FOR CLINICAL TRANSLATION OF BIODEGRADABLE PMO NANOPARTICLES

The study on biodegradable PMO is at a beginning stage and further experiments are needed to be carried out with various animal models. There are two main issues that need further investigation. One issue concerns the mechanism of degradation. While it is clear that these nanoparticles are degraded, it is necessary to follow the process of degradation. Time course studies as well as various detection methods need to be employed to understand how the nanoparticles undergo degradation. Second, it is necessary to examine excretion of the nanoparticles. This can be investigated first using the chicken egg model to see whether the degradation products can be recovered in the waste compartment. Careful studies need to be carried out to examine whether 100% of the nanomaterial will be excreted out. The study can be followed up using mouse models.

5. CHICKEN EGG TUMOR MODEL AS A CONVENIENT AND VERSATILE ANIMAL MODEL TO EVALUATE NANOMATERIALS

The chicken egg tumor model has a number of advantages over other animal models such as mouse xenografts. First, tumor is rapidly formed within 3–4 days in the chicken egg system. With patient tumor, tumor formation occurred in 4 days after transplanting minced tumor samples. This contrasts with relatively long time it takes to form a tumor on xenograft mice (a couple of weeks). The rapid tumor formation in the chicken egg likely represents nutrient rich conditions of the CAM membrane as well as incomplete formation of the immune system in the chicken egg at the time of transplantation [21,22]. Second, the CAM assay is economical, as a fertilized chicken egg costs less than a dollar, while an immune compromised mouse costs almost $100.

Because the CAM assay is relatively easy to set up and also because it is cost effective, the assay can be scaled up to carry out high throughput screening. Thus, the CAM assay can be used to screen a library of chemical compounds to identify novel anticancer drugs. Also, a library of nanoparticles can be screened to identify materials with excellent drug delivery property. Fig. 8 summarizes three applications of the chicken egg tumor model.

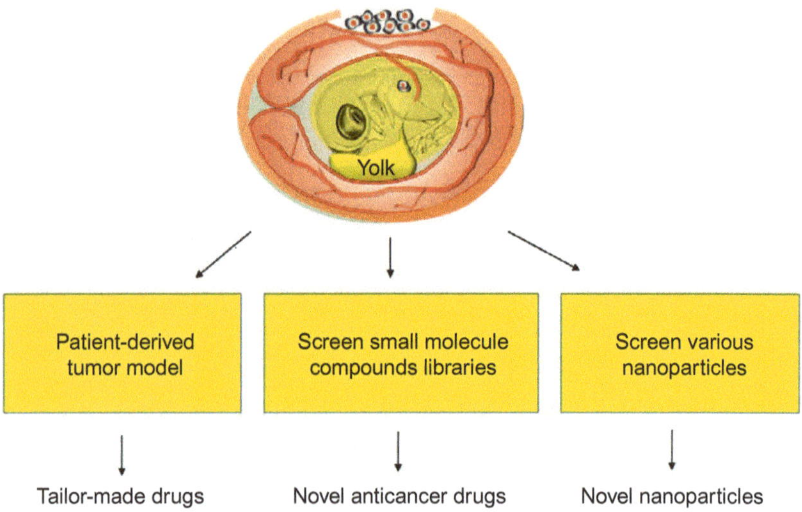

Fig. 8 Potential application of the chicken egg tumor model. First, the chicken egg tumor model can be used to screen various nanoparticles to identify novel nanoparticles effective for tumor targeting and tumor elimination. Second, the model can be used to screen small molecule compound libraries to identify novel anticancer drugs. Third, patient-derived tumor model can be established using the chicken egg model that can help in searching for tailor-made drugs.

6. PERSONALIZED MEDICINE AND THE CHICKEN EGG TUMOR MODEL

Patient-derived tumor models have recently received significant attention among clinicians and scientists engaged in cancer therapy [23–25]. Because the tumor from each individual patient is different, it is necessary to study individual tumor in order to achieve effective therapy. To reproduce tumor from patients, mainly two different approaches have been taken. One is to use tumor spheroids. This involves preparing cancer cells or cancer mass from the patient tumor and then cultures them in media. Significant progress has been made on this approach. However, the method has shortcoming in the sense that the spheroids are quite different from the actual patient tumor. For example, patient tumor can be highly vascularized, but the tumor spheroids are devoid of tumor vessels. The other approach is to use xenograft mice. Patient tumor can be transplanted onto immune compromised mice. This is a good way to reproduce patient tumor. However, this approach requires an extended period to establish a tumor in mice. Thus, the chicken egg tumor model has advantages over these approaches.

We have recently succeeded in transplanting tumor from ovarian cancer patients [11]. This was carried out in collaboration with Carlotta Glackin's group at the City of Hope Cancer Center. Tumor samples removed from ovarian cancer patients (Fig. 9 upper panel) were minced into small fragments. The fragment was placed on top of the CAM membrane and incubated. Four days after the transplantation, tumor mass grew and then continued to enlarge by day 6 (Fig. 9). Histopathological analysis showed that the tumor made on the CAM membrane resembles that of patient tumor. This model thus can be used as a patient-derived tumor model. Ovarian cancer represents a major type of cancer for women [26,27]. After surgery, recurrence of this cancer occurs and there has not been effective treatment for the recurrent ovarian cancer developed. Thus, there is a need for new therapy development.

The chicken egg tumor model has a potential to make significant contribution to personalized medicine [28]. Because of a short period needed to establish patient tumors, it is possible to build up a number of eggs that contain the patient tumor. They can be used to screen various anticancer drugs (Fig. 10). Optimum drug against an individual patient tumor can be identified. The tailor-made drug can be used for the treatment of this particular patient.

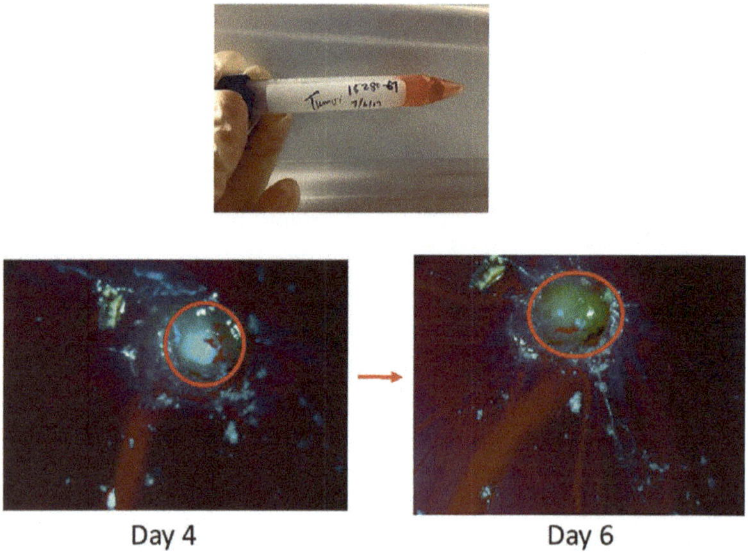

Fig. 9 Patient tumor formed on the CAM membrane. Ovarian tumor sample from the patient (shown in the tube) can be minced and placed on top of the CAM membrane in the chicken egg. Tumor is formed 4 days after the transplantation and the tumor enlarged by day 6. *Adapted from V. Binh, S. Shahin, J. Croissant, Y. Fatieiev, K. Matsumoto, T. Le-Hoang Doan, T. Yik, S. Simargi, A. Conteras, L. Ratliff, C. Mauriello Jimenez, L. Raehm, N. Khashab, J.-O. Durand, C. Glackin, F. Tamanoi, Chick chorioallantoic membrane assay as an in vivo model to study the effect of nanoparticle-based anticancer drugs in ovarian cancer, Sci. Rep. 8 (2018).*

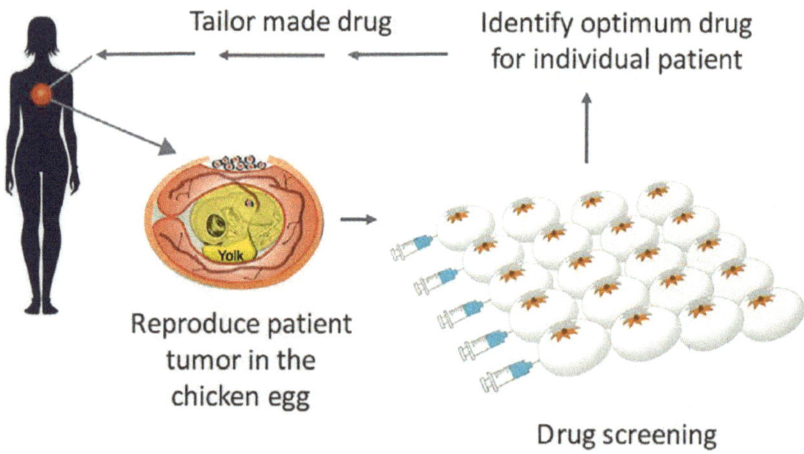

Fig. 10 Tailor-made medicine may be achieved by employing the chicken egg tumor model. In this proposal to realize personalized medicine, patient tumor will be reproduced in the chicken egg and multiple eggs with patient tumor will be prepared. These eggs will be used to screen drugs to identify an optimum drug for an individual patient. This will be a tailor-made drug for this particular patient.

ACKNOWLEDGMENTS

This work was supported by a grant from JSPS KAKENHI grant number JP15K21764 (to F.T.) and from Vietnam National University Ho Chi Minh City grant number VNU-A-562-2018-50-01. We thank Hiroko Imai for her assistance in the preparation of the manuscript.

REFERENCES

[1] R.E. Yanes, F. Tamanoi, Development of mesoporous silica nanomaterials as a vehicle for anticancer drug delivery, Ther. Deliv. 3 (3) (2012) 389–404.
[2] N. Poonia, V. Lather, D. Pandita, Mesoporous silica nanoparticles: a smart nanosystem for management of breast cancer, Drug Discov. Today 23 (2) (2018) 315–332.
[3] A. Watermann, J. Brieger, Mesoporous silica nanoparticles as drug delivery vehicles in cancer, Nanomaterials 7 (7) (2017) 189.
[4] R.R. Castillo, M. Colilla, M. Vallet-Regí, Advances in mesoporous silica-based nanocarriers for co-delivery and combination therapy against cancer, Expert Opin. Drug Deliv. 14 (2) (2017) 229–243.
[5] J. Lu, M. Liong, Z. Li, J.I. Zink, F. Tamanoi, Biocompatibility, biodistribution, and drug-delivery efficiency of mesoporous silica nanoparticles for cancer therapy in animals, Small 6 (16) (2010) 1794–1805.
[6] J. Lu, Z. Li, J.I. Zink, F. Tamanoi, In vivo tumor suppression efficacy of mesoporous silica nanoparticles-based drug-delivery system: enhanced efficacy by folate modification, Nanomed. Nanotechnol. Biol. Med. 8 (2) (2012) 212–220.
[7] J. Croissant, X. Cattoën, M.W.C. Man, A. Gallud, L. Raehm, P. Trens, M. Maynadier, J.-O. Durand, Biodegradable ethylene-bis(propyl)disulfide-based periodic mesoporous organosilica nanorods and nanospheres for efficient in-vitro drug delivery, Adv. Mater. 26 (35) (2014) 6174–6180.
[8] E.A. Prasetyanto, A. Bertucci, D. Septiadi, R. Corradini, P. Castro-Hartmann, L. De Cola, Breakable hybrid organosilica nanocapsules for protein delivery, Angew. Chem. Int. Ed. 55 (10) (2016) 3323–3327.
[9] L. Maggini, I. Cabrera, A. Ruiz-Carretero, E.A. Prasetyanto, E. Robinet, L. De Cola, Breakable mesoporous silica nanoparticles for targeted drug delivery, Nanoscale 8 (13) (2016) 7240–7247.
[10] J.G. Croissant, Y. Fatieiev, K. Julfakyan, J. Lu, A.-H. Emwas, D.H. Anjum, H. Omar, F. Tamanoi, J.I. Zink, N.M. Khashab, Biodegradable oxamide-phenylene-based mesoporous organosilica nanoparticles with unprecedented drug payloads for delivery in cells, Chemistry 22 (42) (2016) 14806–14811.
[11] V. Binh, S. Shahin, J. Croissant, Y. Fatieiev, K. Matsumoto, T. Le-Hoang Doan, T. Yik, S. Simargi, A. Conteras, L. Ratliff, C. Mauriello Jimenez, L. Raehm, N. Khashab, J.-O. Durand, C. Glackin, F. Tamanoi, Chick chorioallantoic membrane assay as an in vivo model to study the effect of nanoparticle-based anticancer drugs in ovarian cancer, Sci. Rep. 8 (2018) 8524.
[12] P. Nowak-Sliwinska, T. Segura, M.L. Iruela-Arispe, The chicken chorioallantoic membrane model in biology, medicine and bioengineering, Angiogenesis 17 (4) (2014) 779–804.
[13] A. Herrmann, D. Moss, V. Sée, The chorioallantoic membrane of the chick embryo to assess tumor formation and metastasis, in: D. Ribatti (Ed.), Tumor Angiogenesis Assays: Methods and Protocols, Springer New York, New York, NY, 2016, pp. 97–105.
[14] M. Klingenberg, J. Becker, S. Eberth, D. Kube, J. Wilting, The chick chorioallantoic membrane as an in vivo xenograft model for Burkitt lymphoma, BMC Cancer 14 (1) (2014) 339.

[15] R. Manjunathan, M. Ragunathan, Chicken chorioallantoic membrane as a reliable model to evaluate osteosarcoma—an experimental approach using SaOS2 cell line, Biol. Proced. Online 17 (1) (2015) 10.
[16] Z. Zuo, T. Syrovets, Y. Wu, S. Hafner, I. Vernikouskaya, W. Liu, G. Ma, T. Weil, T. Simmet, V. Rasche, The CAM cancer xenograft as a model for initial evaluation of MR labelled compounds, Sci. Rep. 7 (2017) 44686.
[17] F. Durupt, D. Koppers-Lalic, B. Balme, L. Budel, O. Terrier, B. Lina, L. Thomas, R.C. Hoeben, M. Rosa-Calatrava, The chicken chorioallantoic membrane tumor assay as model for qualitative testing of oncolytic adenoviruses, Cancer Gene Ther. 19 (2012) 58–68.
[18] A. Vargas, M. Zeisser-Labouèbe, N. Lange, R. Gurny, F. Delie, The chick embryo and its chorioallantoic membrane (CAM) for the in vivo evaluation of drug delivery systems, Adv. Drug Deliv. Rev. 59 (11) (2007) 1162–1176.
[19] H. Maeda, J. Wu, T. Sawa, Y. Matsumura, K. Hori, Tumor vascular permeability and the EPR effect in macromolecular therapeutics: a review, J. Control. Release 65 (1) (2000) 271–284.
[20] S. Wilhelm, A.J. Tavares, Q. Dai, S. Ohta, J. Audet, H. Dvorak, W.C.W. Chan, Analysis of nanoparticle delivery to tumours, Nat. Rev. Mater. 1 (2016) 16014.
[21] B.D. Janković, K. Isaković, M.L. Lukić, N.L. Vujanović, S. Petrović, B.M. Marković, Immunological capacity of the chicken embryo. I. Relationship between the maturation of lymphoid tissues and the occurrence of cell-mediated immunity in the developing chicken embryo, Immunology 29 (3) (1975) 497–508.
[22] D. Ribatti, The chick embryo chorioallantoic membrane in the study of tumor angiogenesis, Rom. J. Morphol. Embryol. 49 (2008) 131–135.
[23] J. Jung, H.S. Seol, S. Chang, The generation and application of patient-derived xenograft model for cancer research, Cancer Res. Treat. 50 (1) (2018) 1–10.
[24] F. Weeber, S.N. Ooft, K.K. Dijkstra, E.E. Voest, Tumor organoids as a pre-clinical cancer model for drug discovery, Cell Chem. Biol. 24 (9) (2017) 1092–1100.
[25] M.-Z. Jin, R.-R. Han, G.-Z. Qiu, X.-C. Ju, G. Lou, W.-L. Jin, Organoids: an intermediate modeling platform in precision oncology, Cancer Lett. 414 (2018) 174–180.
[26] W. Chen, R. Zheng, P.D. Baade, S. Zhang, H. Zeng, F. Bray, A. Jemal, X.Q. Yu, J. He, Cancer statistics in China, 2015, CA Cancer J. Clin. 66 (2) (2016) 115–132.
[27] C. Allemani, H.K. Weir, H. Carreira, R. Harewood, D. Spika, X.-S. Wang, F. Bannon, J.V. Ahn, C.J. Johnson, A. Bonaventure, R. Marcos-Gragera, C. Stiller, G. Azevedo e Silva, W.-Q. Chen, O.J. Ogunbiyi, B. Rachet, M.J. Soeberg, H. You, T. Matsuda, M. Bielska-Lasota, H. Storm, T.C. Tucker, M.P. Coleman, Global surveillance of cancer survival 1995–2009: analysis of individual data for 25 676 887 patients from 279 population-based registries in 67 countries (CONCORD-2), Lancet 385 (9972) (2015) 977–1010.
[28] Y. Wang, Q. Zhao, N. Han, L. Bai, J. Li, J. Liu, E. Che, L. Hu, Q. Zhang, T. Jiang, S. Wang, Mesoporous silica nanoparticles in drug delivery and biomedical applications, Nanomed. Nanotechnol. Biol. Med. 11 (2) (2015) 313–327.

CHAPTER SEVEN

Layered Silicates as a Possible Drug Carrier

Soontaree (Grace) Intasa-ard*, Makoto Ogawa†,[1]
*School of Molecular Science and Engineering, Vidyasirimedhi Institute of Science and Technology (VISTEC), Wangchan, Rayong, Thailand
†School of Energy Science and Engineering, Vidyasirimedhi Institute of Science and Technology (VISTEC), Wangchan, Rayong, Thailand
[1]Corresponding author: e-mail address: waseda.ogawa@gmail.com

Contents

Abstract

The potential of layered silicates as drug carrier is overviewed. Due to their large surface area and expandable interlayer space to accommodate drug molecules, layered silicates have a potential as carrier of various molecules. In addition to the electrostatic interactions between negatively charged layered silicates and positively charged drug molecules, the organic modification of the surface of layered silicates has been applied to accommodate a variety of drug molecules not only cationic ones. The *in vitro* release experiment of the accommodated drug molecules has been reported under the acidic conditions. In order to discuss the future direction of layered silicates as drug carrier, materials' variation of layered silicates and their modification, and the reported stimuli-responsive hybrids based on layered silicates were introduced.

1. INTRODUCTION

In this chapter, layered silicates will be introduced as another candidate of drug carrier as well as other medical application. The required characteristics of drug carrier are to (1) accommodate drug molecules (with larger content), (2) deliver to the target, and (3) release the drug at the site on

The Enzymes, Volume 44
ISSN 1874-6047
https://doi.org/10.1016/bs.enz.2018.08.003
© 2018 Elsevier Inc.
All rights reserved.

demand. One of the basic prerequisites as drug carrier materials is the lack of toxicity. Layered silicates of various structures and composition satisfy the requirements, so that the research on this direction has been done [1–3].

The biomedical application of layered solids (nanosheets) has been reported recently and there are several comprehensive reviews on this aspect [4–7]. The interlayer space of layered solids is often expandable (or swellable) to accommodate guest species in the interlayer spaces and the surface properties are possibly modified by the intercalation [8,9]. The "expandable 2-dimensional nanospace" is regarded as a nanovessel for accommodating various kinds of guest species. The confinement of molecules in nanospace brings peculiar phenomena, for example, dyes in the interlayer space are stabilized [10–13]. Modification of layered solids has been conducted for the uptake of a specific molecule, which cannot be accommodated directly [8,9]. Here, characteristics of the layered silicates and possible materials design based on layered silicates for biomedical used will be summarized followed by the introduction of reported examples of the application of layered silicates for drug delivery.

2. STRUCTURAL CHARACTERISTICS OF LAYERED SILICATES

Among possible layered solids used as scaffolds for materials design, smectites have been used most extensively for a wide range of applications including biomedical ones. Smectites are terms of clay classification meaning of swellable 2:1 type layered silicate (the structure is shown in Fig. 1 together with those of other clay minerals used for the host-guest complexation) consisting of negatively charged silicate layer and the charge compensating interlayer cation [14]. The terms of bentonite and montmorillonite have often been used in a same meaning, while bentonite is a term of natural resource and montmorillonite is a name of a clay mineral. The interlayer cation, which compensates the negative charge of the silicate layer, is exchangeable by the reactions in the suspension and in the solid-state [15–17].

3. SURFACE MODIFICATION

In addition to the structural and compositional variation of smectites (Fig. 1 and Table 1), the surface modification with organic/inorganic cations and polymers is possible (Fig. 2). By utilizing this ability, ionically neutral drug molecules have been incorporated into smectites.

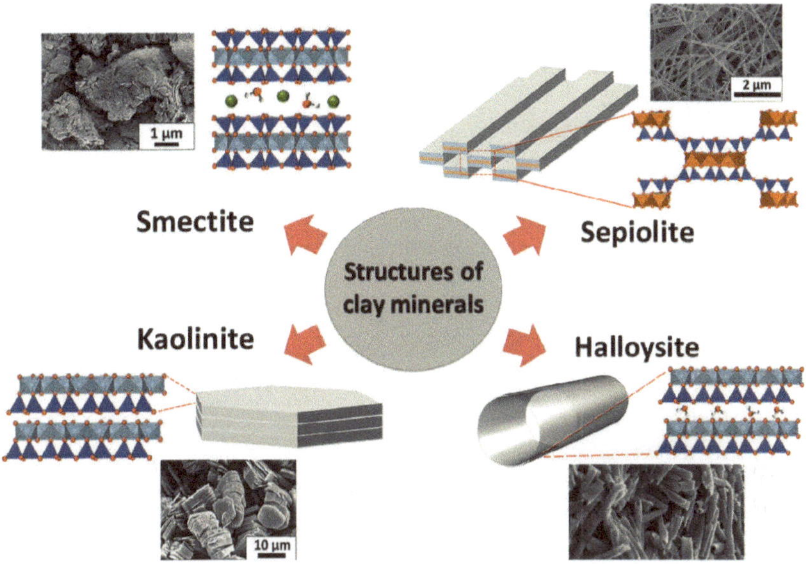

Fig. 1 Structure variation of clay mineral.

Table 1 List of Available Smectites

Type of Octahedral Sheet	Clay	General Composition
Dioctahedral	Montmorillonite	$(M_y^{+} \times nH_2O)\left(Al_{2-y}^{3+}Mg_y^{2+}\right)Si_4^{4+}O_{10}(OH)_2$
	Beidellite	$(M_x^{+} \times nH_2O)Al_2^{3+}(Si_{4-x}^{4+}Al_x^{3+})O_{10}(OH)_2$
	Nontronite	$(M_x^{+} \times nH_2O)Fe_2^{3+}(Si_{4-x}^{4+}Al_x^{3+})O_{10}(OH)_2$
	Volkonskoite	$(M_x^{+} \times nH_2O)Cr_2^{3+}(Si_{4-x}^{4+}Al_x^{3+})O_{10}(OH)_2$
Trioctahedral	Hectorite	$(M_y^{+} \times nH_2O)\left(Mg_{3-y}^{2+}Li_y^{+}\right)Si_4^{4+}O_{10}(OH)_2$
	Saponite	$(M_x^{+} \times nH_2O)Mg_3^{2+}(Si_{4-x}^{4+}Al_x^{3+})O_{10}(OH)_2$
	Sauconite	$(M_x^{+} \times nH_2O)Zn_3^{2+}(Si_{4-x}^{4+}Al_x^{3+})O_{10}(OH)_2$

Long-chain alkylammonium ions have been studied most extensively in the chemistry of organophilic smectites and the practical application of the organophilic smectites as adsorbents [18] is reported. Phospholipids have been utilized for environmentally benign organoclay [19]. Several nonionic surfactants have been used for the surface modification of smectites [20,21].

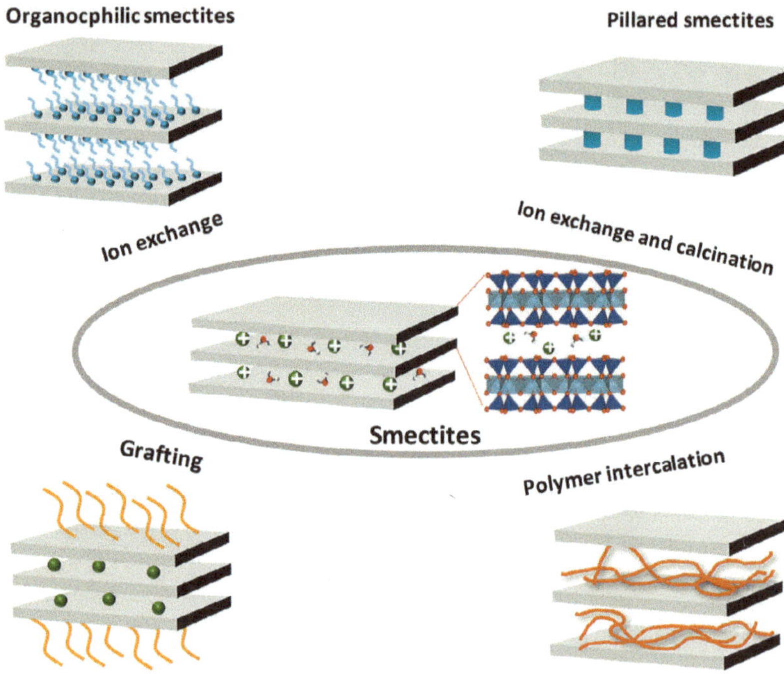

Fig. 2 Schematic drawing of the surface modification of smectites.

Intercalation of alkylammonium ions with more complex structures into layered silicates [22] has been reported in order to design hydrophobic nanospaces more precisely.

Microporous and mesoporous solids have been obtained by crosslinking the nanosheets. The pioneering example is the pillaring with polyoxycations (e.g., $[AlO_4Al_{12}(OH)_{24}(H_2O)_{12}]^{7+}$) [23]. Nanoporous solids composed of silicate layers and metal/metal oxide finite particles have been prepared [24,25]. The microporous solids composed of silicate nanosheet and small organoammonium cations (e.g., TMA) have been prepared and used for the separation/sensing and other functional materials [26–28]. The adsorptive properties of smectites modified with aliphatic and aromatic ammonium ions have also been investigated [29–32]. Pore size and porosity are controlled by selecting pillaring agents. Neostigmine, phenyltrimethylammonium, and 3-(trifluoromethyl)phenyltrimethylammonium were used to find the effects of the surface modification on the adsorption capacity of 2-phenylphenol [33]. In addition, the charge-transfer interactions between adsorbate and adsorbent were a possible driving force for the guest accommodation [34].

The intercalation of water-soluble polymers is a way to modify the surface properties of smectites [35,36]. If compared with the modification of organoammonium ions, the polymer intercalation has advantages; the interlayer expansion can be controlled by the amount of polymer; for the organoammonium case, the amount cannot be used to control the expansion due to the segregation.

4. ACCOMMODATION OF DRUG MOLECULES

Cationic drug molecules have been incorporated into the interlayer space of smectites as summarized in Table 2 [37–50]. Drug molecules with amino group have been accommodated into smectites with electrostatic interactions. The amount of drug molecule on smectites is controlled by the cation exchange capacity of smectites (*ca.* 60–100 meq/100 g smectite). *In vitro* release has been examined under various conditions such as in artificial intestinal juice pH 7, simulated gastric (pH 1.2) condition and so on. Depending on the pH, the release mechanism and the efficiency are different. Under the gastric (pH 1.2) condition, not only the release of the accommodated cationic drug molecules by the exchange with proton but also the decomposition (dissolution) of the silicate layer is possible.

Nonionic and anionic drug molecules have also been combined with smectites, and the interactions between the drug and smectite have been proposed as hydrogen bonding and the ion-dipole interactions between the interlayer cation of smectite and the drug molecule. OH group of smectite is proposed to play a role for the drug accommodation, while OH group is not exposed to the surface of smectite. Further detailed characterization on the interactions and the location of accommodated drug molecules are worth conducting.

5. SURFACE MODIFICATION FOR THE CONTROLLED RELEASE OF DRUG MOLECULES

The surface modification of the drug/smectite has been done to control the release performances (Fig. 3). Table 3 summarizes the examples of the surface modification [51–66]. The basic idea is coating the drug/smectite with polymeric or biocompatible species such as alginate, chitosan and commercial acrylate polymers (Eudragit). The expected function of the coating is targeted release through delayed release, gastro resistance, pH-dependent

Table 2 Examples of Drug Delivery System Based on Clay Minerals

Clay	Drug Molecule	Therapeutic Activity	Proposed Interactions	Release	Ref.
Montmorillonite (MMT)	Promethazine chloride	No info	– Electrostatic interactions between NH^+ group and clay surface – Hydrogen bonding	*In vitro* release in artificial intestinal juice pH 7 ○ 30% (0.394 mol/g) after 360 min	[37]
	Buformin hydrochloride	No info	Electrostatic interactions between NH^+ group and clay surface	*In vitro* release in artificial intestinal juice pH 7 ○ 70% (0.488 mmol/g) after 360 min	[37]
Bentonite ○ Chlorite ○ Kaolinite ○ Illite ○ Montmorillonite	5-Fluorouracil	Colorectal cancer	– Free surface adsorption – Na^+ and OH^- replacement	Not investigated	[38]
Montmorillonite (MMT)	Timolol maleate (TM) (β-adrenergic blocking agent)	Effective against hypertension, arrhythmias, angina pectoris, and secondary prevention of myocardial infraction	Electrostatic interactions between NH^+ group and clay surface	*In vitro* release in simulated gastric (pH 1.2) and intestinal fluids (pH 7.4) ○ pH 1.2: 43% after 9 h ○ pH 7.4: 48% after 9 h	[39]
Montmorillonite (MMT)	Tramadol hydrochloride (TH)	Relieve physical pain caused by surgical operation or chronic diseases	Electrostatic interactions between NH^+ group and clay surface	*In vitro* release in simulated gastric (pH 1.2) and intestinal fluids (pH 7.4) ○ SGF: 72.8% after 120 min ○ SIF: 41.8% after 120 min	[40]

Magnesium aluminum silicate (MAS) (mixture of smectite clays including montmorillonite, saponite)	Propranolol hydrochloride (PPN) (β-adrenoceptor blocking agent)	Use in angina and hypertension	– Electrostatic interactions between NH^+ group and clay surface – Hydrogen bonding	*In vitro* release in 0.1 N HCl and pH 6.8 PBS o 0.1 N HCl: 89% after 7 h o buffer pH 6.8: 74% after 7 h	[41]
Laponite XLG (gel forming grade of laponite)	Tetracycline (antimicrobial agent)	Treat periodontal disease	Electrostatic interactions between NH^+ group and clay surface	*In vitro* release in saliva-like solution (SLS) at 37 °C o SLS (pH 5.2): 52.3% after 72 h o SLS (pH 7.2): 15.4% after 72 h	[42]
Laponite (LAP)	Doxorubicin (DOX)	Antitumor	Hydrogen bonding or electrostatic interactions	*In vitro* release in pH 7.4 PBS and pH 5.4 acetic acid sodium acetate buffer solution o pH 5.4: 29.5% after 9 days o pH 7.4: 10.8% after 9 days	[43]
Fluorohectorite (synthetic hectorite)	Ciprofloxacin (Cipro)	Antibiotic	Electrostatic interactions between negative charge COO^- and cation in the interlayer space	*In vitro* release in SGF at 37 and 65 °C (temperature-controlled release systems) o 37 °C: 10% after 10 h o 65 °C: 25% after 10 h	[44]
Montmorillonite (MMT)	Chlorpromazine (CPZ) (antipsychotic drug)	Psychosis treatment for patient with severe dementia	– Electrostatic interactions between NH^+ group and clay surface – Hydrogen bonding between CPZ and water molecule in the interlayer space	*In vitro* release in SGF (pH 1.2, oral administration) o gradual fast release within 250 min, then reach 75% after 780 min	[45]

Continued

Table 2 Examples of Drug Delivery System Based on Clay Minerals—cont'd

Clay	Drug Molecule	Therapeutic Activity	Proposed Interactions	Release	Ref.
Smectite (mixture of montmorillonite and saponite)	Theophylline (TP)	Bronchodilator in the treatment of asthma and chronic obstructive pulmonary disease	Electrostatic interactions between NH^+ group and clay surface	*In vitro* release in PBS (pH 7.4), SGF (pH 1.2), and SIF (pH 6.8 and 7.4) ∘ SGF: no drug release ∘ SIF (both pH): ~40% after 2 h	[46]
Montmorillonite (MMT)	Ibuprofen (IBU)	Treat rheumatoid arthritis, osteoarthritis, moderate pain	Electrostatic interactions between COO^- groups with OH^- groups of clay	*In vitro* release in HCl (pH 1.2) and, phosphate buffer (pH 7.4) at 37°C ∘ pH 1.2: 11.0% after 120 min ∘ pH 7.4: 30.7% after 120 min	[47]
Bentonite (mainly montmorillonite) ∘ Ca^{2+} bentonite ∘ Zn^{2+} bentonite	Naproxen	Anti-inflammatory drug use for reduction of moderate to severe pain, fever, inflammation and stiffness	Intercalated as neutral molecule displacing water and binding with clay via C=O group to the Zn^{2+} through direct linkage or water bridge	*In vitro* release in PBS solution at pH 7.4 ∘ ~80% after 240 min in 1.0 M NaCl medium	[48]
Unmodified montmorillonite (Ca^{2+}, Mg^{2+}, Cu^{2+}, Fe^{3+}, Al^{3+})	Gallic acid (antioxidant)	Skin protection and skin-lightening	Intercalated as neutral molecule displacing the water and binding with clay via C=O group to the cation through direct linkage or water bridge	*In vitro* release in PBS solution (pH 3 and 7) ∘ 70% after 1 h	[49]
Laponite	Dexamethasone(DEX)	Treat posterior segment ocular diseases	Hydrogen bonding between C=O and OH of drug and clay	*In vitro* release in saline solution ∘ 38% after 10 times of extraction	[50]

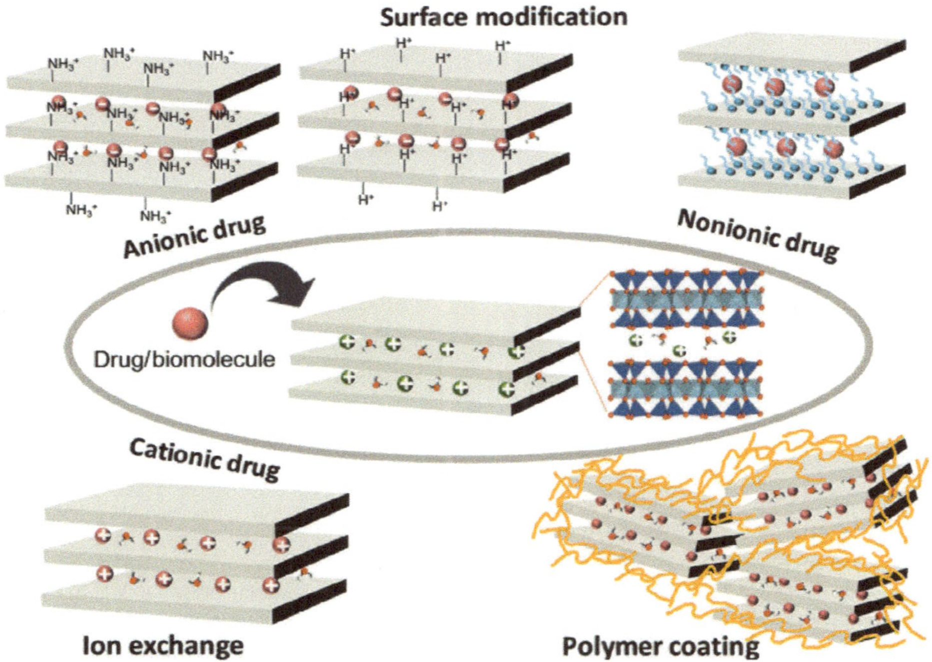

Fig. 3 Schematic drawing for the drug incorporation in smectites and the surface modification for the controlled release.

Table 3 Examples of the Modification of Clay-Drugs for Controlled Release

Clay	Drug	Modification of the Hybrid	Proposed Interactions	Drug Release	Ref.
Montmorillonite (MMT), saponite (SA), laponite (LA)	Donepezil	Coated with Eudragit® E-100	Electrostatic interactions between NH^+ and OH^- of clay double layer arrangement of drug in MMT compared to SA and LA (monolayer) lead to larger encapsulation amount (33%)	*In vitro* release in simulated human gastric media (pH 1.2) ○ release from clay without coating depend on CEC ○ larger CEC gave stronger interactions and sustained release (7%, 15%, 37% from MMT, SA, LA after 3 h) ○ Eudragit® E-100 accelerate release rate by ion exchange of NH_4^+ from Eudragit® E-100 with intercalated drug, and improve the hydration of the hybrid (release rate increase from 12% to 43% (MMT), 23% to 83% (SA), and 61% to 83% (LA) after 3 h)	[51]
Montmorillonite (MMT)	Ofloxacin (OFL)	Entrapped in chitosan microsphere	Electrostatic interactions between NH^+ in OFL and MMT	*In vitro* release in SGF (pH 1.2) and PBS (pH 7.4) ○ release rate is slower at pH 1.2 than that at pH 7.4 ○ 50% of drug was depleted at pH 1.2, while 71% at pH 7.4 after 4 h	[52]

Montmorillonite (MMT)	Ranitidine hydrochloride (RT)	Coated with Eudragit® E-100	Electrostatic interactions between RT and MMT	*In vitro* release in SGF (pH 1.2) 51% of RT was released after 12h ○ releasing amount became higher (64%, 85%, and 90%) by utilizing Eudragit® E-100	[53]
Montmorillonite (MMT)	Buspirone hydrochloride (BUH)	Coated with Eudragit® L 100 55	Electrostatic interactions between BUH and MMT	*In vitro* release in buffer (pH 6.8) ○ without coating: BUH release 23–29% after 12h ○ coating: BUH release 54% after 12h	[54]
Montmorillonite (MMT)	Procainamide hydrochloride (PA)	Compounded with alginate (AL) and coated with chitosan	– Electrostatic interactions between PA and MMT – Hydrogen bonding between AL and MMT	*In vitro* release in buffer (pH 1.2 and 7.4) ○ pH 1.2: PA-MMT-AL-CS release 11% after 2h (less than 38% of PA-MMT) due to the change of AL to water insoluble alginic acid ○ pH 7.4: PA-MMT-AL-CS release 25% after 70h	[55]
Magnesium aluminum silicate (MAS) (mixture of montmorillonite and saponite)	Propranolol HCl	Compounded with alginate (SA)	Electrostatic interactions between drug and MAS	*In vitro* release in HCl 0.1M and PBS (pH 6.8) ○ 46.2% after 10h (higher amount of MAS lead to lower amount of release drug) ○ 46.2% after 10h (higher concentration of $CaCl_2$ to form gel lead to lower amount of drug release)	[56]

Continued

Table 3 Examples of the Modification of Clay-Drugs for Controlled Release—cont'd

Clay	Drug	Modification of the Hybrid	Proposed Interactions	Drug Release	Ref.
Mesoporous synthetic hectorites (MSH) (hectorite and organohectorite)	Quinine (QUI)	Coated with sodium alginate (AL)	Electrostatic interactions between QUI and MSH	*In vitro* release in 0.1 HCl (pH 1.2) and PBS (pH 7.4) ○ without coating: 63% and 94% release from MSH1 and MSH2 after 8 h ○ coating: 25% release for MSH1–QUI/AL and 45% for MSH2–QUI/AL release after 8 h ○ alginate coating led a slow release in the gastric environment	[57]
Montmorillonite	Glutathione (GSH)	Coated with polyvinylacetaldiethylamino-acetate (AEA)	Electrostatic interactions between GSH and MMT	*In vivo* antioxidant activity	[58]
Organomodified laponite (CTAB)	Doxorubicin	Coated with poly(lactide-*co*-glycolide) (PLGA) forming poly(lactide-*co*-glycolide) (PLGA)-laponite-F68 nanocomposite (PNC) vesicle	Not mentioned (adhering to the inner walls of the dried vesicles)	– *In vitro* release in PBS (pH 7.4) ○ 50% after 70 min – *In vitro* cytotoxicity by MTT assay	[59]
Montmorillonite (cloisite Na^+)	Quinine HCl	Coated with Eudragit® L 100 and chitosan (CS)	– Electrostatic interactions between NH_3^+ of drug and OH^- of clay – Electrostatic interactions between clay and NH_2 group of chitosan	*In vitro* release in HCl 0.1 M (pH 1.2) and PBS (pH 7.4) ○ 86–97% release of drug depending on the amount of CS (weaker interactions of drug and clay in the presence of CS) ○ polymer coating prevented the release (dissolution in acidic condition)	[60]

Montmorillonite (MMT)	5-Fluorouracil (5-FU)	Compounded with alginate (Alg) and coated with chitosan (CS)	Electrostatic interactions of 5-FU and MMT and polymer	*In vitro* release in HCl 0.1 M (pH 1.2) and PBS (pH 7.4) ○ pH 7.4 showed faster release than that at pH 1.2	[61]
Montmorillonite (MMT)	Capecitabine (Xeloda)	Entrapped in poly (L-lactide) (PLLA) to form microcomposite sphere (MPs)	Electrostatic interactions of xeloda and MMT and polymer	*In vitro* release in PBS (pH 7.4) ○ CAP/MMT: 24% release after 20 h and 46% after 72 h ○ CAP/MPs: 12% release after 20 h and 17% after 72 h ○ MPs showed slower release due to the low permeability of the water in an interior of MPs	[62]
Montmorillonite (MMT)	Nortriptyline (NT)	Compounded with poly (L-lactide) (PLLA) to form microcomposite sphere (MPs)	Electrostatic interactions between NT and MMT	*In vitro* release in PBS (pH 7.4) ○ NT/MMT: 30% release after 10 h and 42% after 72 h ○ NT/MPs: 8% release after 10 h and 14% after 72 h The slower release of drug from MPs due to the low permeability of water inside	[63]
	Venlafaxine (VFX)	Compounded with poly (L-lactide) (PLLA) to form microcomposite sphere (MPs)	Electrostatic interactions between VFX and MMT	*In vitro* release in PBS (pH 7.4) ○ VFX/MMT: 32–37% release after 10 h and 38–42% release after 48 h ○ VFX/MPs: 5–6% release after 10 h and 8–9% after 48 h	[63]

Continued

Table 3 Examples of the Modification of Clay-Drugs for Controlled Release—cont'd

Clay	Drug	Modification of the Hybrid	Proposed Interactions	Drug Release	Ref.
Sepiolite (SV) (fibrous structure)	Oxaprozin (OXA) (poorly water-soluble drug)	Cogrinding with cyclodextrin (randomly-methylated beta-CD (RAMEB))	Not mentioned	*In vitro* drug release in buffer solution of pH 5.5	[64]
Montmorillonite (MMT)	Mesalazine	Compounded with alginate	Electrostatic interactions between drug and MMT	*In vitro* release in buffer (pH 1.2) and PBS (pH 7.4) ○ 5% at pH 1.2 and 20.7% at pH 7.4	[65]
Montmorillonite (MMT)	Olanzapine (OLZ)	Coated with mixture of alginate (ALG) and xanthan gum (XG) biopolymer	– Electrostatic interactions between OLZ and MMT – Electrostatic interactions between biopolymer with MMT	*In vitro* release in SGF (pH 1.2, 6.8, and 7.4) ○ Selective release at pH 7.4 due to the decrease in the protonated form OLZ at higher pH	[66]

release, and so on. *In vitro* release has been examined for the surface modified drug/smectite under various conditions.

As discussed in chapter "Drug delivery and bone infection" by Vallet-Regí et al. of the present review on surface modification, the modification of interlayer space is possible to accommodate molecules with biological and medical activities, the modification of interlayer space is possible to accommodate molecules with biological and medical activities. The modification of smectites by protonation [67], exchange with alkylammonium ions [68], exchange with phosphatidyl choline [69], and adsorption of poly(ethylene-oxides) [70] has been reported to control uptake, storage, release and response of the hybrids.

6. SUMMARY AND FUTURE PERSPECTIVES

As summarized in Tables 2 and 3, there are several examples of the accommodation of drug molecules on layered silicates and the surface coating to control the drug release. *In vitro* release experiments have been done so far to show the potential and limitation of layered silicates as drug carrier. Optimizing the release performances (such as rate and quantity, as well as stimuli response) has been investigated in the applications not only for drug delivery but also for flavor and agriculture applications. Smectite has been used for slow-release formulations of pesticide and flavor molecules; it was reported that slow-release of alachlor inhibited the growth of green foxtail [71,72]. The desorption of flavor molecule (hinokitiol) from hinokitiol-poly (ethylene terephthalate)-TMA-montmorillonite (Kunipia F) was suppressed due to the entrapment of hinokitiol in TMA-montmorillonite [73].

Stimuli-response is another key, and responses to photoirradiation, temperature, magnetic field, and so on have been proposed for the drug delivery. As an example, photo-responsive adsorption/desorption has been achieved in layered silicate systems where photochromic moieties were immobilized in the interlayer space [74,75].

Particle size is another key issue for the drug delivery. As summarized in Tables 2 and 3, various natural and synthetic layered silicates have been used so far. In general, natural smectites (montmorillonite or purified bentonite) contain micrometer sized fraction, while synthetic smectites (i.e., laponite and sumecton) are finite particles with the size of several tens of nm. In addition, the syntheses of smectite are also possible toward desired composition and particle size for the specific target delivery [76,77].

Efforts have been made for the detailed understanding and design on biological and chemical aspects of the application of layered nanomaterials for the drug delivery. Physicochemical characteristics of the drug-clay systems such as rheology and photostability have been studied so far [78,79]. Cytotoxicity of modified and unmodified clays has been examined already [80,81]. Further study on the preparation of layered silicates with desired particle size, the accommodation of drug molecules, the surface modification for the targeted delivery and controlled release, and imparting stimuli-responsive function for the system are worth conducting to realize practical delivery systems.

ACKNOWLEDGMENTS

This work was supported by the Research Chair Grant 2017 (grant number FDA-CO-2560-5655) from the National Science and Technology Development Agency (NSTDA), Thailand. One of the authors (S.G.I.) acknowledges Vidyasirimedhi Institute of Science and Technology for the scholarship to her PhD study.

REFERENCES

[1] E. Ruiz-Hitzky, P. Aranda, M. Darder, M. Ogawa, Hybrid and biohybrid silicate based materials: molecular versus block-assembling bottom-up processes, Chem. Soc. Rev. 40 (2011) 801–828.
[2] S.M. Auerbach, K.A. Carrado, P.K. Dutta, Handbook of Layered Materials, Marcel Dekker, New York, 2004.
[3] F. Bergaya, B.K.G. Theng, G. Lagaly, Handbook of Clay Science, Elsevier Science, Amsterdam, 2006.
[4] J.M. Oh, T.T. Biswick, J.H. Choy, Layered nanomaterials for green materials, J. Mater. Chem. 19 (2009) 2553–2563.
[5] E. Ruiz-Hitzky, P. Aranda, M. Darder, G. Rytwo, Hybrid materials based on clays for environmental and biomedical applications, J. Mater. Chem. 20 (2010) 9306–9321.
[6] J.I. Dawson, R.O.C. Oreffo, Clay: new opportunities for tissue regeneration and biomaterial design, Adv. Mater. 25 (2013) 4069–4086.
[7] L.A.d.S. Rodrigues, A. Figueiras, F. Veiga, R.M. de Freitas, L.C.C. Nunes, E.C. da Silva Filho, C.M. da Silva Leite, The systems containing clays and clay minerals from modified drug release: a review, Colloids Surf. B Biointerfaces 103 (2013) 642–651.
[8] T. Okada, Y. Ide, M. Ogawa, Organic-inorganic hybrids based on ultrathin oxide layers—designed nanostructures for molecular recognition, Chem. Asian J. 7 (2012) 1980–1992.
[9] T. Okada, Y. Seki, M. Ogawa, Designed nanostructures of clay for controlled adsorption of organic compounds, J. Nanosci. Nanotechnol. 14 (2014) 2121–2134.
[10] Y. Kodera, H. Kageyama, H. Sekine, Y. Inada, Photo-stable chlorophylls conjugated with montmorillonite, Biotechnol. Lett. 14 (1992) 119.
[11] M. Ogawa, M. Sohmiya, Y. Watase, Stabilization of photosensitizing dyes by the complexation with clay, Chem. Commun. 47 (2011) 8602–8604.
[12] T. Goto, M. Ogawa, Efficient photocatalytic oxidation of benzene to phenol by metal complex-clay/TiO_2 hybrid photocatalyst, RSC Adv. 6 (28) (2016) 23794–23797.

[13] T. Goto, M. Ogawa, Visible light responsive photocatalytic flow reactor composed of titania film photosensitized by metal complex-clay hybrid, ACS Appl. Mater. Interfaces 7 (23) (2015) 12631–12634.
[14] B.K.G. Theng, The Chemistry of Clay-Organic Reactions, Adam Hilger, London, 1974.
[15] M. Ogawa, T. Handa, K. Kuroda, C. Kato, Formation of organoammonium-montmorillonites by solid-solid reactions, Chem. Lett. (1990) 71–74.
[16] M. Ogawa, Y. Nagafusa, K. Kuroda, C. Kato, Solid-state intercalation of acrylamide into smectites and Na-taeniolite, Appl. Clay Sci. 7 (1992) 291–302.
[17] S.G. Intasa-ard, K.J. Imwiset, S. Bureekaew, M. Ogawa, Mechanochemical way for the preparation of intercalation compounds, Dalton Trans. 47 (2018) 2896–2916.
[18] M. Ogawa, K. Kuroda, Preparation of inorganic-organic nanocomposites through intercalation of organoammonium ions into layered silicates, Bull. Chem. Soc. Jpn. 70 (1997) 2593–2618.
[19] B. Wicklein, M. Darder, P. Aranda, E. Ruiz-Hitzky, Bio-organoclays based on phospholipids as immobilization hosts for biological species, Langmuir 26 (7) (2010) 5217–5225.
[20] M. Ogawa, N. Kanaoka, K. Kuroda, Preparation of smectite-dodecyldimethylamine-n-oxide intercalation compounds, Langmuir 14 (1998) 6969–6973.
[21] R. Guégan, E. Veron, L. Le Forestier, M. Ogawa, S. Cadars, Structure and dynamics of nonionic surfactant aggregates in layered materials, Langmuir 33 (38) (2017) 9759–9771.
[22] M. Ogawa, D. Iwata, Arrangements of interlayer quaternary ammonium ions in a layered silicate, octosilicate, Cryst. Growth Des. 10 (2010) 2068–2072.
[23] T.J. Pinnavaia, Intercalated clay catalysts, Science 220 (4595) (1983) 365–371.
[24] A. Galarneau, A. Barodawalla, T.J. Pinnavaia, Porous clay heterostructures formed by gallery-templated synthesis, Nature 374 (6522) (1995) 529–531.
[25] P.B. Malla, P. Ravindranathan, S. Komarneni, R. Roy, Intercalation of copper metal clusters in montmorillonite, Nature 351 (6327) (1991) 555–557.
[26] M. Ogawa, M. Takahashi, C. Kato, K. Kuroda, Oriented microporous film of tetramethylammonium saponite, J. Mater. Chem. 4 (1994) 519–523.
[27] M. Ogawa, M. Takahashi, K. Kuroda, Intercalation of p-nitroaniline into tetramethylammonium-saponite film under electric field and its optical second harmonic generation, Chem. Mater. 6 (1994) 715–717.
[28] M. Ogawa, T. Handa, K. Kuroda, C. Kato, T. Tani, Photochemical hole burning of 1,4-dihydroxyanthraquinone intercalated in a pillared layered clay mineral, J. Phys. Chem. 96 (1992) 8116–8119.
[29] S.A. Boyd, S. Shaobai, J.F. Lee, M.M. Mortland, Pentachlorophenol sorption by organo-clays, Clays Clay Miner. 36 (2) (1988) 125–130.
[30] Y. Yan, T. Bein, Molecular recognition through intercalation chemistry: immobilization of organoclays on piezoelectric devices, Chem. Mater. 5 (7) (1993) 905–907.
[31] J.A. Smith, P.R. Jaffe, Comparison of tetrachloromethane sorption to an alkylammonium-clay and an alkyldiammonium-clay, Environ. Sci. Technol. 25 (12) (1991) 2054–2058.
[32] S. Nir, T. Undabeytia, D.Y. Marcovich, Y. El-Nahhal, T. Polubesova, C. Serban, G. Rytwo, G. Lagaly, B. Rubinet, Optimization of adsorption of hydrophobic herbicides on montmorillonite preadsorbed by monovalent organic cations: interaction between phenyl rings, Environ. Sci. Technol. 34 (7) (2000) 1269–1274.
[33] Y. Seki, M. Ogawa, The removal of 2-phenylphenol from aqueous solution by adsorption onto organoclays, Bull. Chem. Soc. Jpn. 83 (2010) 712–715.
[34] T. Okada, M. Ogawa, 1,1′-Dimethyl-4,4′-bipyridinium-smectites as a novel adsorbent of phenols from water through charge-transfer interactions, Chem. Commun. (2003) 1378–1379.

[35] M. Ogawa, M. Tsujimura, K. Kuroda, Incorporation of tris(2,2′-bipyridine)ruthenium(ii) in a synthetic swelling mica with poly(vinylpyrrolidone), Langmuir 16 (2000) 4202–4206.
[36] M. Sohmiya, S. Omata, M. Ogawa, Two dimensional size controlled confinement of poly(vinyl pyrrolidone) in the interlayer space of swelling clay mineral, Polym. Chem. 3 (2012) 1069–1074.
[37] I. Fejér, M. Kata, I. Erös, O. Berkesi, I. Dékány, Release of cationic drugs from loaded clay minerals, Colloid Polym. Sci. 279 (2001) 1177–1182.
[38] F.H. Lin, Y.H. Lee, C.H. Jian, J.M. Wong, M.J. Shieh, C.Y. Wang, A study of purified montmorillonite intercalated with 5-fluorouracil as drug carrier, Biomaterials 23 (2002) 1981–1987.
[39] G.V. Joshi, B.D. Kevadiya, H.A. Patel, H.C. Bajaj, R.V. Jasra, Montmorillonite as a drug delivery system: intercalation and in vitro release of timolol maleate, Int. J. Pharm. 374 (2009) 53–57.
[40] Y. Chen, A. Zhou, B. Liu, J. Liang, Tramadol hydrochloride/montmorillonite composite: preparation and controlled drug release, Appl. Clay Sci. 49 (2010) 108–112.
[41] S. Rojtanatanya, T. Pongjanyakul, Propranolol–magnesium aluminum silicate complex dispersions and particles: characterization and factors influencing drug release, Int. J. Pharm. 383 (2010) 106–115.
[42] M. Ghadiri, H. Hau, W. Chrzanowski, H. Agus, R. Rohanizadeh, Laponite clay as a carrier for in situ delivery of tetracycline, RSC Adv. 3 (2013) 20193–20201.
[43] S. Wang, Y. Wu, R. Guo, Y. Huang, S. Wen, M. Shen, J. Wang, X. Shi, Laponite nanodisks as an efficient platform for doxorubicin delivery to cancer cells, Langmuir 29 (2013) 5030–5036.
[44] A. Rivera, L. Valdés, J. Jiménez, I. Pérez, A. Lam, E. Altshuler, L.C. de Ménorval, J.O. Fossum, E.L. Hansen, Z. Rozynek, Smectite as ciprofloxacin delivery system: intercalation and temperature-controlled release properties, Appl. Clay Sci. 124–125 (2016) 150–156.
[45] M.A. Djebbi, S. Boubakri, Z. Bouaziz, M.S. Elayachi, P. Namour, N. Jaffrezic-Renault, A. Ben Haj Amara, Extended-release of chlorpromazine intercalated into montmorillonite clays, Microporous Mesoporous Mater. 267 (2018) 43–52.
[46] V. Trivedi, U. Nandi, M. Maniruzzaman, N.J. Coleman, Intercalated theophylline-smectite hybrid for pH-mediated delivery. Drug Deliv. Transl. Res. (2018) 1–9, https://doi.org/10.1007/s13346-018-0478-8.
[47] J.P. Zheng, L. Luan, H.Y. Wang, L.F. Xi, K.D. Yao, Study on ibuprofen/montmorillonite intercalation composites as drug release system, Appl. Clay Sci. 36 (2007) 297–301.
[48] A. Tabak, N. Yilmaz, E. Eren, B. Caglar, B. Afsin, A. Sarihan, Structural analysis of naproxen-intercalated bentonite (Unye), Chem. Eng. J. 174 (2011) 281–288.
[49] M. Rabiei, H. Sabahi, A.H. Rezayan, Gallic acid-loaded montmorillonite nanostructure as a new controlled release system, Appl. Clay Sci. 119 (2016) 236–242.
[50] J.M. Fraile, E. Garcia-Martin, C. Gil, J.A. Mayoral, L.E. Pablo, V. Polo, E. Prieto, E. Vispe, Laponite as carrier for controlled in vitro delivery of dexamethasone in vitreous humor models, Eur. J. Pharm. Biopharm. 108 (2016) 83–90.
[51] J.K. Park, Y.B. Choy, J.M. Oh, J.Y. Kim, S.J. Hwang, J.H. Choy, Controlled release of donepezil intercalated in smectite clays, Int. J. Pharm. 359 (2008) 198–204.
[52] S. Hua, H. Yang, A. Wang, A pH-sensitive nanocomposite microsphere based on chitosan and montmorillonite with in vitro reduction of the burst release effect, Drug Dev. Ind. Pharm. 36 (2010) 1106–1114.
[53] G.V. Joshi, B.D. Kevadiya, H.C. Bajaj, Controlled release formulation of ranitidine-containing montmorillonite and Eudragit® E-100, Drug Dev. Ind. Pharm. 36 (2010) 1046–1053.

[54] G.V. Joshi, B.D. Kevadiya, H.C. Bajaj, Design and evaluation of controlled drug delivery system of buspirone using inorganic layered clay mineral, Microporous Mesoporous Mater. 132 (2010) 526–530.
[55] B.D. Kevadiya, G.V. Joshi, H.C. Bajaj, Layered bionanocomposites as carrier for procainamide, Int. J. Pharm. 388 (2010) 280–286.
[56] T. Pongjanyakul, T. Rongthong, Enhanced entrapment efficiency and modulated drug release of alginate beads loaded with drug–clay intercalated complexes as microreservoirs, Carbohydr. Polym. 81 (2010) 409–419.
[57] G.V. Joshi, R.R. Pawar, B.D. Kevadiya, H.C. Bajaj, Mesoporous synthetic hectorites: a versatile layered host with drug delivery application, Microporous Mesoporous Mater. 142 (2011) 542–548.
[58] M. Baek, J.H. Choy, S.J. Choi, Montmorillonite intercalated with glutathione for antioxidant delivery: synthesis, characterization, and bioavailability evaluation, Int. J. Pharm. 425 (2012) 29–34.
[59] B.P. Nair, C.P. Sharma, Poly(lactide-co-glycolide)–laponite–F68 nanocomposite vesicles through a single-step double-emulsion method for the controlled release of doxorubicin, Langmuir 28 (2012) 4559–4564.
[60] G.V. Joshi, B.D. Kevadiya, H.M. Mody, H.C. Bajaj, Confinement and controlled release of quinine on chitosan–montmorillonite bionanocomposites, J. Polym. Sci. A Polym. Chem. 50 (2012) 423–430.
[61] F. Farshi Azhar, A. Olad, A study on sustained release formulations for oral delivery of 5-fluorouracil based on alginate–chitosan/montmorillonite nanocomposite systems, Appl. Clay Sci. 101 (2014) 288–296.
[62] B.D. Kevadiya, S.S. Chettiar, S. Rajkumar, H.C. Bajaj, H. Brahmbhatt, J.C. Chaudhari, R.P. Thumbar, D. Jhala, M.V. Rao, Evaluation of montmorillonite/poly (L-lactide) microcomposite spheres as ambidextrous reservoirs for controlled release of Capecitabine (Xeloda) and assessment of cell cytotoxic and oxidative stress markers, Compos. Sci. Technol. 90 (2014) 193–201.
[63] S. Rajkumar, B.D. Kevadiya, H.C. Bajaj, Montmorillonite/poly (L-lactide) microcomposite spheres as reservoirs of antidepressant drugs and their controlled release property, Asian J. Pharm. Sci. 10 (2015) 452–458.
[64] P. Mura, F. Maestrelli, C. Aguzzi, C. Viseras, Hybrid systems based on "drug—in cyclodextrin—in nanoclays" for improving oxaprozin dissolution properties, Int. J. Pharm. 509 (2016) 8–15.
[65] H.J. Hong, J. Kim, Y.J. Suh, D. Kim, K.M. Roh, I. Kang, pH-sensitive mesalazine carrier for colon-targeted drug delivery: a two-fold composition of mesalazine with a clay and alginate, Macromol. Res. 25 (2017) 1145–1152.
[66] A.S. Oliveira, A.C.S. Alcântara, S.B.C. Pergher, Bionanocomposite systems based on montmorillonite and biopolymers for the controlled release of olanzapine, Mater. Sci. Eng. C 75 (2017) 1250–1258.
[67] F. Ibrahim, O. Sivak, C. Wong, P. Hopkins, A. Midha, J. Gordon, J.W. Darlington, K.M. Wasan, Impact of co-administration of protonated nanostructured aluminum silicate (cholesterol absorption inhibitor) on the absorption of lipid soluble vitamins D3 and K1: an assessment of pharmacokinetic and in vitro intraluminal processing, Eur. J. Pharm. Sci. 49 (2013) 125–132.
[68] S.H. Hsu, H.J. Tseng, H.S. Hung, M.C. Wang, C.H. Hung, P.R. Li, J.J. Lin, Antimicrobial activities and cellular responses to natural silicate clays and derivatives modified by cationic alkylamine salts, ACS Appl. Mater. Interfaces 1 (2009) 2556–2564.
[69] N. Zhou, S. Fang, D. Xu, J. Zhang, H. Mo, J. Shen, Montmorillonite–phosphatidyl choline/PDMS films: a novel antithrombogenic material, Appl. Clay Sci. 46 (2009) 401–403.

[70] L. Zampori, P.G. Stampino, C. Cristiani, P. Cazzola, G. Dotelli, Intercalation of poly (ethylene-oxides) in montmorillonite: tailor-made nanocontainers for drug delivery systems, Appl. Clay Sci. 50 (2010) 266–270.
[71] Y. El-Nahhal, S. Nir, T. Polubesova, L. Margulies, B. Rubin, Leaching, phytotoxicity, and weed control of new formulations of alachlor, J. Agric. Food Chem. 46 (1998) 3305–3313.
[72] Y. El-Nahhal, T. Undabeytia, T. Polubesova, Y.G. Mishael, S. Nir, B. Rubin, Organo-clay formulations of pesticides: reduced leaching and photodegradation, Appl. Clay Sci. 18 (2001) 309–326.
[73] R. Ishii, Y. Imai, M. Wada, T. Ebina, T. Hanaoka, F. Mizukami, Adsorption and desorption behaviors of flavor molecules into a microporous pillared clay mineral and the application to flavor capsule composites, Appl. Clay Sci. 33 (2006) 99–108.
[74] T. Okada, M. Sohmiya, M. Ogawa, Photochromic intercalation compounds, in: D.M.P. Mingos (Ed.), Structure and Bonding, Photofunctional Layered Materials, Springer, Oxford, United Kingdom, 2015, pp. 177–212.
[75] T. Okada, N. Nozaki, J. Seo, J.E. Kwon, S.Y. Park, H. Hashizume, M. Ogawa, Photoinduced structural changes of cationic azo dyes confined in a two dimensional nanospace by two different mechanisms, RSC Adv. 7 (2017) 8077–8081.
[76] M. Ogawa, T. Matsutomo, T. Okada, Preparation of hectorite-like swelling silicate with controlled layer charge density, J. Ceram. Soc. Jpn. 116 (2008) 1309–1313.
[77] M. Ogawa, T. Matsutomo, T. Okada, Preparation of iron containing hectorite-like swelling silicate, Bull. Chem. Soc. Jpn. 82 (2009) 408–412.
[78] E. Mascheroni, P. Chalier, N. Gontard, E. Gastaldi, Designing of a wheat gluten/ montmorillonite based system as carvacrol carrier: rheological and structural properties, Food Hydrocoll. 24 (2010) 406–413.
[79] V. Ambrogi, L. Latterini, M. Nocchetti, C. Pagano, M. Ricci, Montmorillonite as an agent for drug photostability, J. Mater. Chem. 22 (2012) 22743–22749.
[80] S. Lordan, J.E. Kennedy, C.L. Higginbotham, Cytotoxic effects induced by unmodified and organically modified nanoclays in the human hepatic HepG2 cell line, J. Appl. Toxicol. 31 (2011) 27–35.
[81] Q. Liu, Y. Liu, S. Xiang, X. Mo, S. Su, J. Zhang, Apoptosis and cytotoxicity of oligo(styrene-co-acrylonitrile)-modified montmorillonite, Appl. Clay Sci. 51 (2011) 214–219.

CHAPTER EIGHT

Organosilica Nanoparticles and Medical Imaging

Michihiro Nakamura[1]
Department of Organ Anatomy & Nanomedicine, Yamaguchi University Graduate School of Medicine, Ube, Japan
[1]Corresponding author: e-mail address: nakam@yamaguchi-u.ac.jp

Contents

Abstract

Medical imaging technology using nanoparticles has several advantages from it varies functional properties. As we described previous chapters, mesoporous silica nanoparticles demonstrated great contribution for nanomedicine progress and it has been expected to cause an innovation in medical field. Recently we developed a novel type of silica nanoparticles, organosilica nanoparticles. Organosilica nanoparticles are both structurally and functionally different from common silica nanoparticles by including mesoporous silica nanoparticles. The organosilica nanoparticles are inherent organic–inorganic hybrid nanomaterials. The interior and exterior functionalities of organosilica nanoparticles are effective for their internal and surface functionalization. Medical imaging using organosilica nanoparticles is making a new field of nano-medical imaging. Multifunctionalizations peculiar to organosilica nanoparticles enable to construct novel medical imaging system. In this chapter we will introduce organosilica nanoparticles, and its applications on advanced medical imaging.

The Enzymes, Volume 44
ISSN 1874-6047
https://doi.org/10.1016/bs.enz.2018.08.002
© 2018 Elsevier Inc.
All rights reserved.

1. INTRODUCTION

Nanomedicine is an interdisciplinary research area that fuses various scientific fields including nanotechnology, organic chemistry, physics, materials science, pharmacology, molecular biology, and biotechnology for medical development [1–20]. The nanoparticles are becoming essential components of medical research and applications. Recent researches revealed that nanoparticles have excellent potentials to create innovative medicine. Nanoparticles have been anticipated as not only improvement of small molecules but also as innovative tools to direct a course of the development of novel medical system. Applications of nanoparticles for biomedical field have various advantages as compared with small molecules such as common drugs and imaging agents. Advantages of nanoparticles were summarized as below.

(a) Numerous superiority: Single molecule has one action, but single nanoparticle has numerous actions. Nanoparticles have their large interior capacity and surface area. Their larger interior capacity makes it available to embody large amounts of imaging agent or drug molecules. Also their larger surface area can bind many specific targeting moieties such as antibodies and ligands, and can show avidity effect. The nanoparticles have numerous actions and can exhibit greater effect than that of one molecule.

(b) Multifunction: Single small molecules have one function. Nanoparticles can contain more than two kinds of functional molecule and nanomaterial, and reveal more than two functions. The nanoparticles containing two kinds of imaging agent such as fluorescent molecules and iron oxide can be applied to multimodal imaging. The nanoparticles containing imaging and therapeutic agents will be applied as novel medicinal agents for both diagnosis and therapy (Theranostics).

(c) Nano-sized effect: Nanoparticles have completely different behavior in body from small molecule. Small molecules have short-term retention in body, and can distribute various organs, tissues, cells, and intracellular organelle. On the other hand, nanoparticles have relative long-term retention. The nanoparticles cannot pass through cell membrane. The nanoparticles are uptaken by macrophage due to their sizes, and distribute the reticuloendothelial system mainly. The enhanced permeability and retention (EPR) effect is well known as an important feature of the nanoparticles for drug delivery [21–25]. The EPR effect is a property that the appropriate sizes of nanoparticles tend to accumulate in tumor tissue much more than that of normal tissues. Because the nanoparticles extravasate into the tumor stroma through the wide fenestrations of the angiogenic vasculature. In addition, tumor tissues usually lack normal lymphatic drainage. The nanoparticles have various nano-sized effects to biodistribution. And the nanoparticles would bring efficient targeting in vivo due to nano-sized effects.

(d) Additive/multiplier effect: The nanoparticles have advantages such as numerous superiority, multifunction, and nano-sized effect as described above. The combination of these advantages produces additive effect as well as multiplier effect. Many targeting moieties on the surface of particles (multiplicity) with more than one targeting molecule (multifunction) can greatly enhance target binding and specificity compared to single molecules due to avidity and multivalent effects.

However, the applications of nanoparticles have various advantages, the challenges in optimizing design of nanoparticles to develop innovative nanomedicine still remain. An understanding of the biodistribution, the interaction with biological system, safety and toxicity of nanoparticles are not good enough even now. Various functional nanoparticles such as liposome, dendrimer, gold nanoparticles, and iron oxide nanoparticles have been developed and investigated in medical imaging. Recently medical imaging of silica nanoparticles also progresses extensively. Unlike quantum dots or iron oxide nanoparticles, silica nanoparticles do not possess an inherent signal that can be exploited for sensitive imaging applications. However, silica nanoparticles can be effectively functionalized for various medical imaging. In addition, various sizes of silica nanoparticles with narrow size distribution can be synthesized. These characteristics of silica nanoparticles have high potential to exhibit advantages of nanoparticles for medical imaging. Therefore, silica nanoparticles can be expected to contribute to biomedical development toward theranostics.

2. SILICA NANOPARTICLES AND PREPARATION

Silica, silicon dioxide (SiO_2), is one of oxide form of silicon (Si). Silicon oxide forms, such as silicate (SiO_4) and silica, are the main constituents of sand and quartz contributing to 60% of the Earth's crust. Si is used as a substrate for integrated circuit chips found in computers, cell phones and other electronic devices. In the food field, silica serves as a preservative and thinning agent. Si based products are also widely utilized for medical imaging. As an example, Si and silicon oxide are used for dietary supplements, catheters and implants, dental fillers and contact lenses [26,28–30]. Silicon, silicon oxide and their products are very familiar to our life.

Silicon and oxygen are main component of silica nanoparticles. In 1968, Stöber et al. reported a synthesis of silica nanoparticles from aqueous alcohol solutions [31]. In this synthesis, silica nanoparticles are formed through the hydrolysis and condensation of alkoxysilane (Fig. 1). The alkoxysilanes contain carbon, but the formed silica nanoparticles do not contain any carbons due to hydrolysis. Therefore, these silica nanoparticles can be termed "inorganosilica nanoparticles." In addition to inorganosilica nanoparticles, another kind of silica nanoparticles has been prepared and reported. These novel silica nanoparticles contain exposed interior and surface carbons as well as organic and functional groups as prepared. Organosilica nanoparticles contain interior and surface functional resides, as well as carbonaceous moieties, as prepared. Various types of novel inorganosilica and organosilica nanoparticles have been developed and applied to biomedical researches recently.

2.1 Inorganosilica Nanoparticle

Traditionally, inorganosilica nanoparticles have been prepared from tetraethoxyorthosilicate (TEOS) or tetramethoxyorthosilicate (TMOS). As prepared, inorganosilica nanoparticles are composed of a simple silica framework

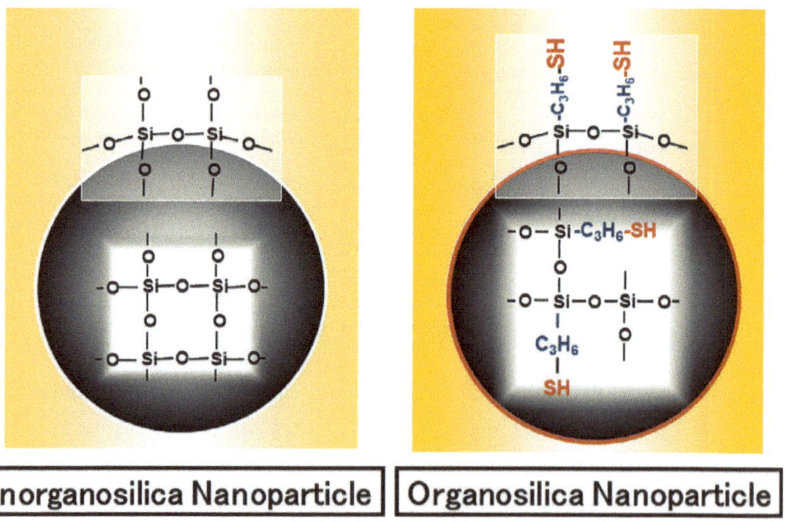

Fig. 1 Schematic structure of organosilica nanoparticles.

(–O–Si–O–) internally, with silanol groups on surface of the particles (Fig. 1). Based on structural feature there are two types of inorganosilica nanoparticles, dense inorganosilica nanoparticles and mesoporous inorganosilica nanoparticles. Dense inorganosilica nanoparticles have been prepared by means of the Stöber method [31] and by reverse microemulsion [32–35].

2.2 Organosilica Nanoparticle

Recently, novel type of silica nanoparticles has been prepared using organosilane successfully. The organosilane has chemical structure that at least one alkoxy chain of the tetraalkoxysilate has been replaced by a hydrocarbon chain (Fig. 1). The organosilane has been used for surface modification (silanization) of inorganosilica nanoparticles [36], but not been used as single silica source to form particles. I have discovered that organosilane can form nanoparticles as single source. The as-prepared organosilica nanoparticles have various organic and functional groups inside of and on the surface of particles, depending on the organosilica source. These organic and functional groups are very useful for multifunctionalizations of the nanoparticles.

One of organosilica nanoparticles, thiol-organosilica nanoparticles, are made from 3-mercaptopropyltrimethoxysilane (MPMS), 3-mercaptopropyltriethoxysilane (MPES), and 3-mercaptopropylmethyldimethoxysilane (MPDMS) as the single source using one-pot syntheses [37–39] (Fig. 1). The syntheses of thiol-organosilica nanoparticles made of three kinds of thiol-organosilicate were different in nanoparticle formation due to the kinds and concentrations of thiol-organosilicate and due to synthetic conditions [38]. The zeta potentials of thiol-organosilica nanoparticles are more negative than those of TEOS nanoparticles [37–39]. The mean diameters of nanoparticles increase with increasing concentrations of silicate, but the size distributions of nanoparticles were dependent on synthetic conditions and silicate sources. As prepared, thiol-organosilica nanoparticles contain interior and surface carbon chains as well as interior and surface thiol residues, as shown schematically in Fig. 1. Size-controlled thiol-organosilica nanoparticles are made from MPMS by means of one-pot syntheses without alcohol and at high temperature (100 °C) [39]. Incubation of the reaction mixtures at 100 °C accelerated the formation of the particles and improved size distribution. The coefficient of variations (CVs) of size distributions were <20%, which was better than the CVs for thiol-organosilica particles synthesized at room temperature. The diameters of the particles formed also depended on the concentrations of MPMS present in the synthetic mixtures.

Recently we report an efficient synthetic method of thiol-organosilica nanoparticles with smaller diameter using an anionic surfactant [40]. Compared with the conventional synthetic methods, >10-fold miniaturization of thiol-organosilica nanoparticles and an approximately 40-fold increase in the production efficiency of smaller thiol-organosilica nanoparticles were achieved using the sodium dodecyl sulfate (SDS)-addition synthetic method or sodium dodecylbenzenesulfonate (SDBS)-addition synthetic method. The CVs of the size distribution of thiol-organosilica NPs synthesized with SDS were sufficiently small. The polydispersity of MPMS NPs synthesized with SDS from DLS measurements was slightly high (27.3%), but the polydispersities of MPES NPs synthesized with SDS and MPDMS NPs synthesized with SDS were 9.6% and 15.7%, respectively. The SDS-addition synthetic method or SDBS-addition synthetic method will be useful for large scale preparation to apply the biomedical applications of thiol-organosilica nanoparticles.

Another organosilica nanoparticles, epoxy-organosilica nanoparticles, are made from 2-(3, 4-epoxycyclohexyl) ethyltrimethoxysilane (EpoMS) [41]. As prepared, the zeta potentials of epoxy-organosilica nanoparticles are positive. The mean diameters and size distributions of the particles also depended on the EpoMS concentration. However, typical Stöber method could not form epoxy-organosilica nanoparticles, and synthesis mixture without alcohol could form epoxy-organosilica nanoparticles with narrow size distribution.

3. FUNCTIONALIZATION OF ORGANOSILICA NANOPARTICLES

The functionalization of nanoparticles is very important for medical imaging. The functionalization of nanoparticles can be classified as internal functionalization and as surface functionalization. Internal functionalization of silica nanoparticles with functional molecules, such as fluorescent molecules, nanomaterials, and drugs, is essential for bioassays, imaging and drug delivery. Internally functionalized silica nanoparticles can provide strong and stable signals for bioassay and imaging, and great pharmacological effect for drug delivery. Surface functionalization of silica nanoparticles by conjugation with various biomolecules, such as proteins, enzymes, peptides, and DNA, is a key technology to improve specificity and affinity to bind target molecule. Optimal fusion of surface and internal functionalization of nanoparticles can realize useful multifunctional nanoparticles for innovative medical imaging. The type and extent of possible multifunctionalization of nanoparticles are largely controlled by the surface and internal properties of the as-prepared nanoparticles. Organosilica nanoparticles are becoming more attractive core particles for medical imaging because they have useful functionalities inside and on surface as prepared. This section describes internal and surface functionalization of organosilica nanoparticles.

3.1 Internal Functionalization

Fluorescence-based detection and imaging techniques have been widely used in recent biochemical research. Primarily fluorescent molecules have been used in fluorescence-based techniques. Traditional fluorescent molecules such as fluorescein, rhodamine, and Cy5 have long been used. In most cases, only one or a few fluorescent molecules can signal one biomolecule recognition event, and typically, only a limited number of fluorescent molecules can be attached to a biomolecule without interfering with its binding specificity. And their low fluorescence intensity and low photostability make them unsuitable for high-sensitive detection and long-time monitoring. Recently, new fluorescent nanoparticles including quantum dots, fluorescent latex nanoparticles, and fluorescent silica nanoparticles have been developed and have shown great utilities for fluorescence applications.

Fluorescent silica nanoparticles have been developed using internal functionalization with fluorescent molecules. During particle synthesis fluorescent molecules can be incorporated into the particle by using two methods, direct method and conjugation method. Direct method has one-step procedure. Fluorescent molecules without any modification were added to synthetic mixture of silica nanoparticles and were incorporated into particles due to interactions between fluorescent molecules and internal structure of silica nanoparticles [37–51]. Conjugation method has at least two-step procedure. First, fluorescent molecule is conjugated with silicate to form silica-fluorescent molecule conjugate. Second, the silica-fluorescent molecules conjugate is mixed with a synthetic mixture of silica nanoparticles to incorporate silica-fluorescent molecules conjugate to silica framework [52–58]. Another type of fluorescent silica nanoparticles, core–shell fluorescent silica nanoparticles, were also reported [59].

3.1.1 Direct Method

Direct method enables simple and low cost preparation of fluorescent silica nanoparticles. The internal functionalization using direct method is dependent on internal properties of the nanoparticles and the synthesis process.

Various inorganosilica nanoparticles containing fluorescent molecules were prepared and applied to bioassay and biomedical imaging [43–51]. Organosilica nanoparticles also contained fluorescent molecules without cross-linking reagents [37–42].

Fluorescent organosilica nanoparticles have been prepared from a one-pot synthesis with direct method. Rhodamine B or Rubpy was contained into thiol-organosilica nanoparticles as well as epoxy-organosilica nanoparticles successfully [37–42]. To compare the incorporation efficacy, direct method was applied to inorganosilica nanoparticles with rhodamine B, but no fluorescence is observed from the inorganic nanoparticles prepared from TEOS. The incorporation efficacy of organosilica nanoparticles differs from that of inorganosilica nanoparticles predominantly [38]. Because organosilica nanoparticles have unique internal structures such as carbonaceous moieties and functional resides such thiol residue, fluorescent molecules can interact with internal structure via hydrophobic interaction and electrostatic interaction. To evaluate ability of direct method, fluorescent-tuned thiol-organosilica nanoparticles were prepared using direct method with rhodamine B and conjugation method with maleimide-rhodamine red conjugates. Fluorescent organosilica nanoparticles can via direct method possess higher fluorescence intensities than those via conjugation method [38]. These reports demonstrated that incorporation of fluorescent molecules into organosilica nanoparticles produced high quantity and quality packaging of functional molecules. Synthesis of organosilica nanoparticles containing functional molecules has high possibilities to create high functional nanomaterials due to not only multiplicity but also additive effect/multiplier effect.

We reported property of fluorescence organosilica nanoparticles containing rhodamine B as a fluorescent imaging probe in vitro [60]. Fluorescent thiol-organosilica nanoparticle with 100 nm diameter was monodisperse upon electron microscopy (Fig. 2A), and the mean diameter and coefficient of variation (%CV) of F-thiol-OS-100 were 101 nm and 9.4%, respectively. The hydrodynamic size of particle was examined using dynamic light scattering (Fig. 2B). The diameter and %CV were 169.8 nm and 9.9%, respectively, and the diameter was higher and %CV were similar as compared with those of transmission electron microscopy due to physical principle. Comparative single particle analysis of quantum dots and fluorescence organosilica nanoparticles revealed difference of each property as fluorescent nanoparticles. Using conventional fluorescence microscopy, we measured the time-dependent changes of fluorescent intensity of single particles of thiol-organosilica nanoparticle containing rhodamine B and of

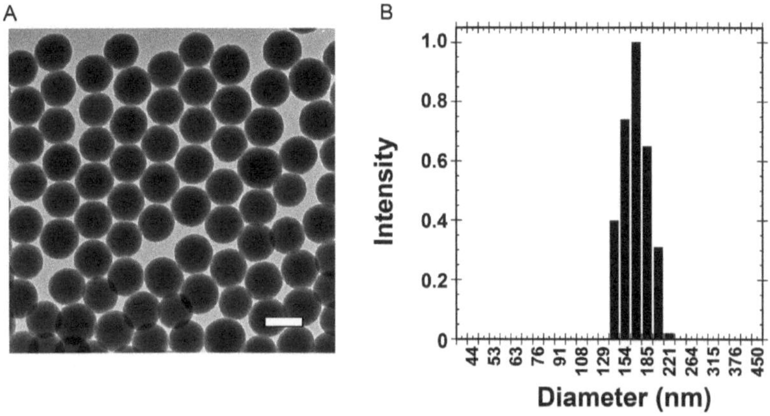

Fig. 2 Characterization of fluorescent organosilica nanoparticles. (A) Transmission electron micrographs of fluorescent organosilica nanoparticles. Scale bar is 100 nm. (B) Dynamic light scattering analysis of fluorescent organosilica nanoparticles. *From M. Nakamura, K. Miyamoto, K. Hayashi, A. Awaad, M. Ochiai, K. Ishimura, Time-lapse fluorescence imaging and quantitative single cell and endosomal analysis of peritoneal macrophages using fluorescent organosilica nanoparticles, Nanomedicine 9 (2013) 274–283.*

quantum dots. The curves of the fluorescent intensity of single particles of 10 fluorescent thiol-organosilica nanoparticles showed similar patterns and weak photobleaching (Fig. 3A). A comparison of 10 quantum dots showed various patterns due to photoblinking (Fig. 3B). The minimum and maximum of the mean of the signal of each quantum dot were 507.4 and 2616.3 A.U., respectively, and minimum and maximum %CV were 16.6% and 135.5%, respectively. To evaluate their quantification abilities, we calculated the mean fluorescent intensities of 10 particles of fluorescent thiol-organosilica nanoparticle and quantum dot, and also the fluorescent intensity per area of rhodamine B. The normalized curves of the mean fluorescent thiol-organosilica nanoparticle fluorescent intensity showed a linear decrease; the intensity remained above 60% of their initial values after 150 cycles (Fig. 3C). The mean intensity at cycle 0 and the %CV of each cycle were 26,104 and 6.3%, respectively. The rhodamine B showed high photobleaching; its intensity declined to <10% of the initial value after 150 cycles (Fig. 3C); the mean value at cycle 0 was 179.2 A.U. per area. The quantum dots did not show clear photobleaching, but variable changes of the mean of fluorescent intensity due to photoblinking (Fig. 3C); the mean and the %CV of each cycle were 3020 A.U. and 55.2%, respectively. Fluorescent intensity between each of the quantum dots showed marked differences, even though these quantum dots were measured on the same glass slide at the same time. The mean of each cycle of quantum dots showed unpredictable and

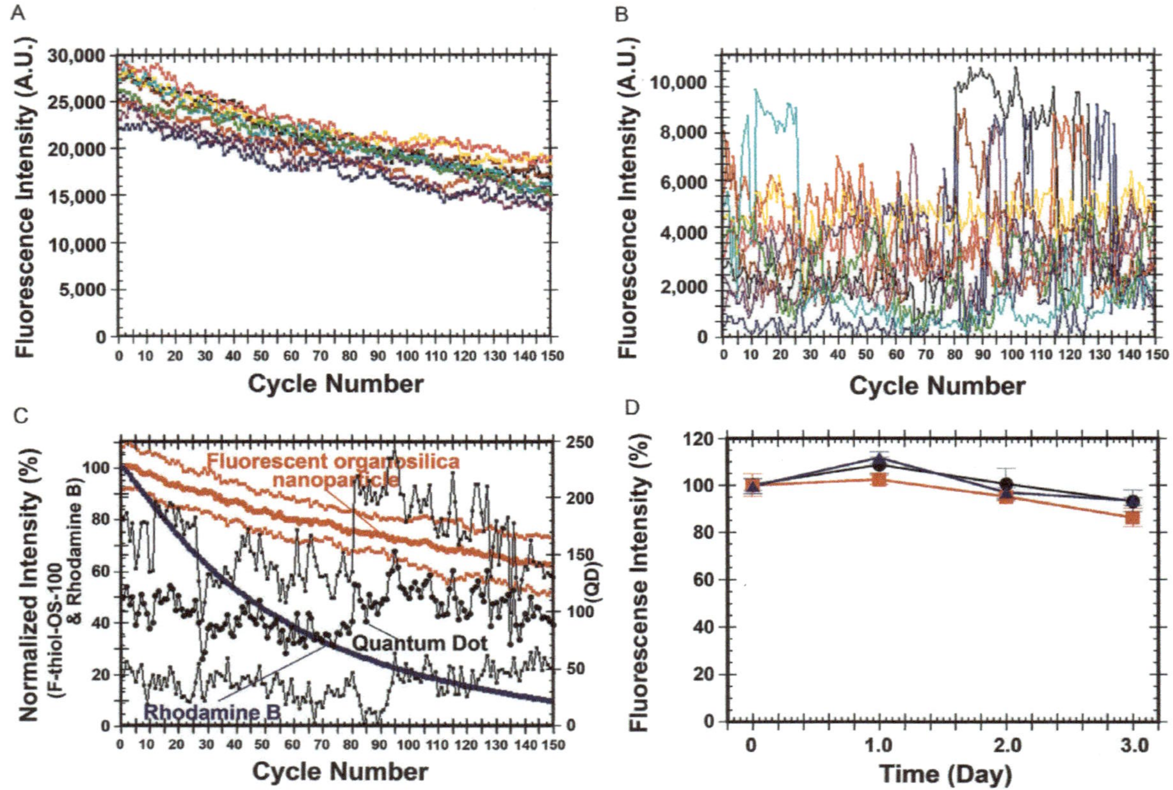

Fig. 3 See legend on opposite page.

pronounced large changes as compared with those of fluorescent thiol-organosilica nanoparticle. Thus, the use of quantum dots for quantitative analysis is compromised due to their photoblinking. Although fluorescent thiol-organosilica nanoparticle displayed photobleaching under higher emission light for single quantum dot analysis using conventional microscopy, the decrease of intensity was linear and predictable (0.26% decrease per cycle). Therefore, it is possible to compensate the loss of fluorescent intensity as a function of cycle.

Stabilities of fluorescent intensities of fluorescent thiol-organosilica nanoparticles containing rhodamine B in acidic condition were evaluated because intracellular environments such as a late-stage endosome and lysosome were acidic (pH <5). As shown in Fig. 3D, fluorescent thiol-organosilica nanoparticle in phthalate solution pH 4 could keep the fluorescent intensity at the same level of those of water and phosphate buffered saline pH 7.4 for 3 days. These results indicated that fluorescent thiol-organosilica nanoparticle did not release fluorescent molecules nor self-quench in the endosomes. As related results, we reported that organosilica particles containing fluorescent molecules could keep their fluorescent intensities in mouse gastrointestinal tract even passed through stomach (pH <2) [61].

Fig. 3 Comparison of photophysical properties of fluorescent organosilica nanoparticles, quantum dots, and rhodamine B molecules. Fluorescence microscopic evaluation of the fluorescent intensities of 10 single fluorescent organosilica nanoparticle (A) and 10 of single quantum dot (B) over 150 cycles (1 cycle = 100 ms). (C) Fluorescence microscopic evaluation of the quantitative ability of fluorescent organosilica nanoparticles (red circles and red solid line), quantum dots (black circles and black solid line), and rhodamine B (blue solid line). The mean (large circle) and standard deviation (small circle) of fluorescent intensity of 10 particles, and the concentration of fluorescent intensity of rhodamine B were calculated. The fluorescent intensities were normalized with respect to the fluorescent intensities at cycle 0 for fluorescent organosilica nanoparticles and rhodamine B, and with a mean intensity for the quantum dots. The normalized values and SD of fluorescent organosilica nanoparticles and rhodamine B appear on the left *y*-axis although those of quantum dots appear on the right *y*-axis. (D) Restoration of fluorescent intensity of fluorescent organosilica nanoparticles. Fluorescent organosilica nanoparticles were incubated in water (black), phosphate buffered saline pH 7.4 (red), or phthalate solution pH 4 (blue). Fluorescent intensities were normalized to those in day 0. Data are presented the mean values and SD for three experiments. *From M. Nakamura, K. Miyamoto, K. Hayashi, A. Awaad, M. Ochiai, K. Ishimura, Time-lapse fluorescence imaging and quantitative single cell and endosomal analysis of peritoneal macrophages using fluorescent organosilica nanoparticles, Nanomedicine 9 (2013) 274–283.*

Thiol-organosilica particles containing propidium iodide (PI) are also prepared using direct method [42]. The PI particles showed unique fluorescent signals as well as X-ray absorption. Orally administered PI particles were observed in the gastrointestinal tract using fluorescent imaging devices and X-ray CT. PI particles showed an alteration of fluorescent properties such as enhancement of the fluorescence intensity and multicolor fluorescence including original fluorescence and near-infrared. An interaction between PI and organosilica framework altered fluorescent properties of PI (Fig. 4). This can be one of multiplier effects of the internal functionalization of organosilica nanoparticles.

3.1.2 Conjugation Method

By means of conjugation method, fluorescent molecules that are difficult to prepare fluorescent silica nanoparticles by direct method can be efficiently incorporated into silica nanoparticles. The typical preparations of fluorescent silica nanoparticles using (3-aminopropyl)triethoxysilane (APS) and amine-reactive fluorescent conjugates are two-step processes. First, APS is attached to isothiocyanate-conjugated fluorescent molecule to form an APS-fluorescent molecule conjugate. Second, the APS-fluorescent molecules conjugate is mixed with a synthetic mixture of silica nanoparticles.

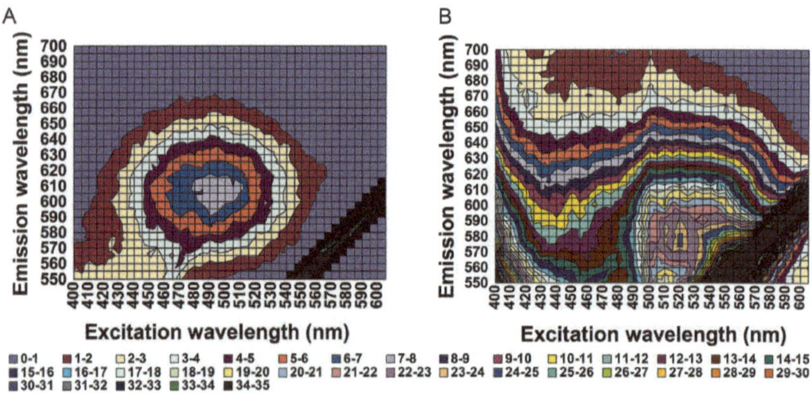

Fig. 4 Fluorescence properties of PI particles. A comprehensive three-dimensional analysis of the excitation, emission, and fluorescence intensities for propidium iodide (PI) (A) and organosilica particles containing PI (B). *From M. Nakamura, A. Awaad, K. Hayashi, K. Ochiai, K. Ishimura, Thiol-organosilica particles internally functionalized with propidium iodide as a multicolor fluorescence and x-ray computed tomography probe and application for non-invasive functional gastrointestinal tract imaging, Chem. Mater. 24 (2012) 3773–3779.*

Fluorescence thiol-organosilica nanoparticles can be prepared using conjugation methods as well as using one-pot syntheses using maleimide-fluorescent molecules conjugates due to the presence of interior thiol residues [38]. Thiol-organosilica nanoparticles retain fluorescent molecules via maleimide-thiol reaction during particle formation. Therefore, preparation of thiol-organosilica nanoparticles using maleimide-fluorescent molecules conjugates is simple and similar process with that of direct method.

3.1.3 Internal Functionalization With Nanomaterials

Nanomaterials have unique functions that are different from functional molecules. Gold nanoparticles have photo-absorbing property due to their surface plasmon and are afforded a cancer therapy strategy known as plasmonic photothermal therapy. Iron oxide nanoparticles with superparamagnetic properties are used as contrast agents in MRI. In addition, iron oxide nanoparticles are also expected therapeutic applications such as magnetic fluid hyperthermia. Internal functionalizations of silica nanoparticles with nanomaterials can be considered various advantages to create useful nanomaterials for biomedical research.

Internal functionalized organosilica nanoparticles with functional nanomaterials were synthesized and reported [62–64]. We reported one-pot synthesis of quantum dot encapsulated with thiol-organosilica layer using MPMS. The thiol-organosilica layer contained fluorescent molecules [62]. The nanoparticles revealed dual fluorescence that was strong fluorescence, high photostability, and reduced photoblinking. There were some additive effects. The quantum dots encapsulated with thiol-organosilica layer could preserve their fluorescent properties under severe acid and alkali conditions due to shield effect of organosilica layer. Furthermore, the nanoparticles showed an extension of the excitation wavelength range for emission of quantum dots. These can be one of the multiplier effects of multifunctionalized organosilica nanoparticles.

Recently we reported the preparation and imaging application of thiol-organosilica nanoparticles containing iron oxide magnetic nanoparticles (MNPs) as the core and rhodamine B in the thiol-organosilica layer (thiol-OS-MNP/Rho) [63] (Fig. 5A). The thiol-OS-MNP/Rho showed enhanced magnetic resonance imaging (MRI) contrast and high fluorescence intensity. The relaxometry of thiol-OS-MNP/Rho revealed a novel coating effect of the organosilica layer to the MNPs. The organosilica layer shortened the T2 relaxation time but not the T1 relaxation time of the MNPs (Fig. 5Ba). The absorbance spectra of thiol-OS-MNP/Rho showed

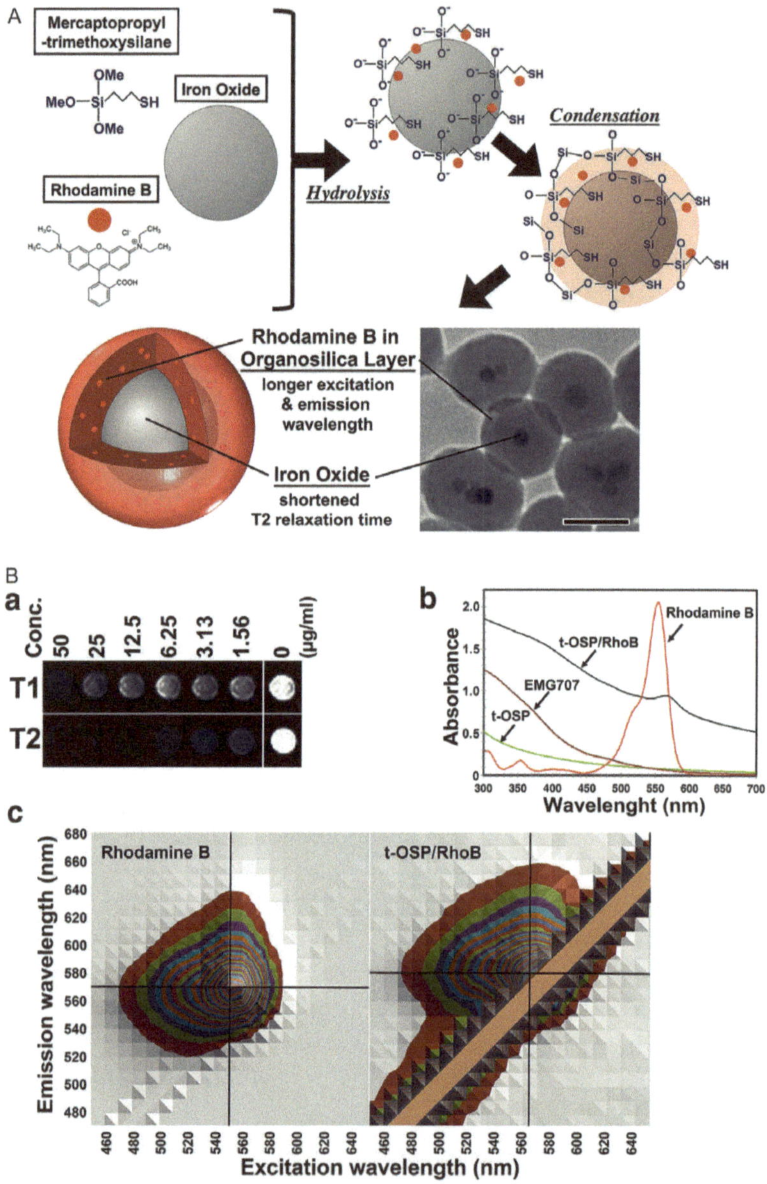

Fig. 5 Organosilica multimodal imaging probe. (A) Scheme of the formation and electron microscopic finding of thiol-OS-MNP/Rho. (B) Panel (a) shows the T1 and T2 weighted images that were obtained with spin echo and turbo spin echo sequence, respectively. Acquisition parameters were as below; (T1 weighted image) TR = 300 ms, TE = 10 ms, 3 mm thickness, FOV = $240 \times 144\,mm^2$, matrix = 256×108, NEX = 8, (T2 weighted image) TR = 5000 ms, TE = 89 ms, 3 mm thickness, FOV = $240 \times 142\,mm^2$, matrix = 448×160, NEX = 4. (b) Absorbance spectra of t-OSP/RhoB, rhodamine B, EMG707, and thiol organosilica particles without rhodamine B nor EMG707. (C) Comprehensive three-dimensional analysis of the excitation, emission, and fluorescence intensities for rhodamine B and thiol-OS-MNP/Rho. *From Ref. M. Nakamura, K. Hayashi, H. Kubo, M. Harada, K. Izumi, Y. Tsuruo, T. Yogo, Mesoscopic multimodal imaging provides new insight to tumor tissue evaluation: an example of macrophage imaging of hepatic tumor using organosilica nanoparticles, Sci. Rep. 7 (1) (2017) 3953.*

a peak at 563 nm, while rhodamine B appears at 554 nm (Fig. 5Bb). In addition, thiol-OS-MNP/Rho showed broad absorbance even above 600 nm. Comprehensive three-dimensional analysis of the excitation, emission, and fluorescence intensities of rhodamine B solution and thiol-OS-MNP/Rho (Fig. 5Bc) was performed. A fluorescence profile of the thiol-OS-MNP/Rho was different from that of rhodamine B solution. The rhodamine B solution had excitation/emission maxima of 550/570 nm. The excitation/emission maxima of the thiol-OS-MNP/Rho were 560/580 nm. These results demonstrated that the shift of the absorbance and fluorescence peaks to longer wavelengths was observed. These results suggested that the fluorescent property of rhodamine B was altered in organosilica layer, and also the multiplier effects of multifunctionalized organosilica nanoparticles.

3.2 Surface Functionalization

The organosilica nanoparticles contain useful reactive groups on their surface as well as internally. Thiol-organosilica nanoparticles have thiol residues on their surfaces as prepared, without any additional procedures. Fourier transform infrared spectroscopy analysis of thiol-organosilica nanoparticles revealed thiol residue derived from mercaptopropyl residue of MPMS [60]. The spectra of MPMS solution, thiol-organosilica nanoparticles, and fluorescent thiol-organosilica nanoparticles containing rhodamine B were presented in Fig. 6. The peaks at 2564 in MPMS solution, 2553 in thiol-organosilica nanoparticles and 2554 cm^{-1} in fluorescent thiol-organosilica nanoparticles originating from the thiol residue can be observed, which is consistent with the fact that the thiol residue with the Si–O–Si framework remains intact after both the hydrolysis and condensation reaction. In parts spectra of thiol-organosilica nanoparticles and fluorescent thiol-organosilica nanoparticles, the formation of the Si–O–Si framework is evidenced by the bands located at 1071 cm^{-1}, and a band centered at 2926–2939 cm^{-1} was assigned to CH_2 units. A broad band centered at 3517 and 3521 cm^{-1} was assigned to Si–OH in fluorescent thiol-organosilica nanoparticles and thiol-organosilica nanoparticles, respectively. Between the spectra of thiol-organosilica nanoparticles and fluorescent thiol-organosilica nanoparticles, there was difference at 1450–1800 cm^{-1} indicating incorporated rhodamine B. These results indicated that the fluorescent thiol-organosilica nanoparticles contained thiol residue derived from mercaptopropyl residue of MPMS and rhodamine B.

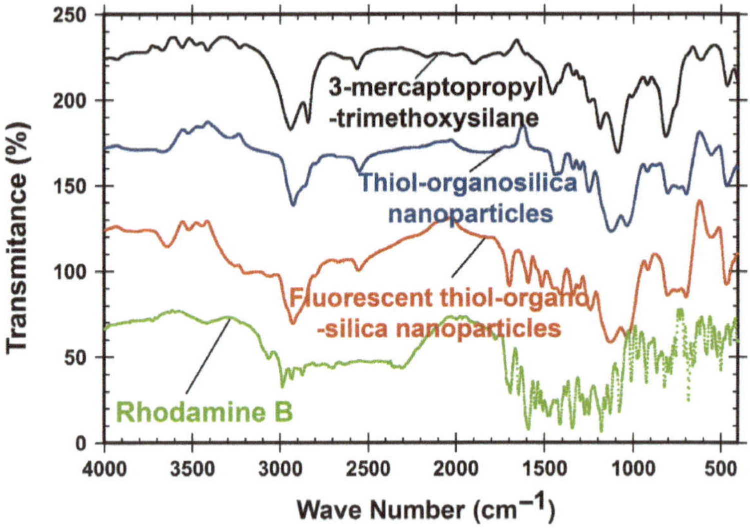

Fig. 6 Fourier transform infrared spectroscopy spectra of 3-mercaptopropyl-trimethoxysilane solution (black), thiol-organosilica nanoparticles (blue), fluorescent thiol-organosilica nanoparticles (red), and rhodamine B (green). *From M. Nakamura, K. Miyamoto, K. Hayashi, A. Awaad, M. Ochiai, K. Ishimura, Time-lapse fluorescence imaging and quantitative single cell and endosomal analysis of peritoneal macrophages using fluorescent organosilica nanoparticles, Nanomedicine 9 (2013) 274–283.*

Thiol residue on thiol-organosilica nanoparticles has specific bindings to thiol residue using thiol/disulfide exchange reactions and chemical coupling reagents such as maleimide. The thiol/disulfide exchange reactions are also useful for conjugating thiol residues via disulfide bonds formed by oxidation of each thiol residue. The disulfide bonds can be controlled by reduction and oxidation. Using thiol residues on the surface, novel nanoparticle-based drug delivery and release systems have been proposed and demonstrated. The disulfide glutathione (GSH) is the most abundant thiol species in cytoplasm, and it could be a major reducing agent [65–67]. Using the difference of concentration between intracellular GSH (1–10 mM) and extracellular GSH (2 μM in plasma), researchers have reported selective intracellular releases of substances. The nanoparticles coated with plasmid DNA via disulfide bonds could release plasmid DNA via dissociation by intracellular disulfide reduction [68]. Thiol-organosilica nanoparticles might be adapted as carriers for nanoparticle-based drug delivery and release systems that use GSH to realize selective intracellular release of drugs and biomolecules.

3.2.1 Adsorption

The surface properties of functionalized thiol-organosilica nanoparticles have been evaluated by various methods such as flow cytometry, fluorescence image analysis, and fluorescence microscopy with inorganosilica nanoparticles as a control [38,39]. To compare the surface properties of thiol-organosilica nanoparticles and inorganosilica nanoparticles, we have performed a dot-blotting procedure [64]. Dot blotting allows for evaluation of the nanoparticles' ability to adsorb proteins on glass slides. In this procedure, the same amount of each nanoparticle was deposited on glass slides and then reacted with a solution containing Cy3-conjugated anti-goat immunoglobulin G (IgG) antibody and examined with a fluorescence image analyzer. The spots of thiol-organosilica nanoparticles showed markedly higher intensity derived from Cy3-conjugated anti-goat IgG as compared with those of inorganosilica nanoparticles, and the fluorescence intensity of thiol-organosilica nanoparticles was about three times that of inorganosilica nanoparticles [37]. These results indicate that thiol-organosilica nanoparticles on glass slides absorbed a greater amount of protein than inorganosilica nanoparticles.

It is well known that the gold nanoparticles have high affinity toward bonding with proteins via thiol functional groups. We also compared the efficacy to adsorb proteins on surface of the thiol-organosilica nanoparticles and that of gold nanoparticles with mean diameter 100 nm by means of flow cytometry [39]. The solutions containing the same number of particles were mixed with various concentrations of GFP solution and analyzed by flow cytometry. The bindings of thiol-organosilica nanoparticles were from 6.3 to 24.5 times those of gold nanoparticles at each concentration of proteins. These results indicated that thiol-organosilica nanoparticles could absorb a greater amount of protein than that of gold nanoparticles.

It is reported that as prepared thiol-organosilica nanoparticles exhibited very good ability to adhere to mucosal surfaces. The disulfide bond formation between thiol groups on the surface of thiol-organosilica nanoparticles and cysteine-rich domains in mucins on mucosal surfaces was proposed as possible mechanism of mucoadhesion. Surface of thiol-organosilica nanoparticles has mucoadhesive properties and is expected for application in bioadhesive mucosal drug delivery systems [69].

Thiol residues react with various functionalities, such as alkyl halides and maleimides, and they easily form covalent bonds with other molecules. Recently, various maleimide-conjugates, including maleimide-conjugated fluorescent molecules, streptavidin, and PEG, have become commercially available. The reaction procedures between maleimide-conjugated molecules and thiol residue of thiol-organosilicate are simple and conjugation efficiencies are high. Specific binding of rhodamine red-maleimide to thiol-organosilica nanoparticles was confirmed with a control experiment using inorganosilica particles [39].

3.2.2 Chemical Conjugation

To bind to target cancer cells, antibody-conjugated thiol-organosilica nanoparticles have been prepared using physical adsorption and chemical bifunctional cross-linking reagent. Anti-human CD20 antibody was physically adsorbed onto the surface of the thiol-organosilica nanoparticles. The CD20 antigen is known to be a target of lymphoma cells, and rituximab, chimeric anti-human CD20 antibody, is approved for treatment of B-cell lymphoma in adults. Antibody adsorbed nanoparticles bound with Raji cells (CD20 positive cells) specifically [39]. Thiol-organosilica nanoparticles reacted with 4-maleimidobutyric acid *N*-succinimidyl ester (NHS) to conjugate antibodies with the surface of the particles via maleimidobutyric acid-thiol linkages. The resulting NHS-surface-modified thiol-organosilica nanoparticles are bound with amino residue of anti-CD20 antibodies covalently. Antibody-conjugated thiol-organosilica nanoparticles have been used for specific binding to CD20 on Raji cells [62].

Thiol-organosilica nanoparticles surface functionalized with PEG were prepared using methoxy poly(ethylene glycol) maleimide (PEG-maleimide) to apply medical imaging of stealth function of PEG [70]. The fluorescent organosilica nanoparticles containing fluorescein (Flu) were PEGylated using methoxy-PEG-maleimide with a molecular weight of 30 kDa in a one-step process (Flu-PEG30K) (Fig. 7). To evaluate the density of PEGs on particle, we performed thermal gravimetric analysis (TGA) of Flu, Flu-PEG12K, and Flu-PEG30K. The initial weight loss under 200 °C is due to the desorption of water molecules from the surface of the particle. The remaining weights were 60.4%, 57.9%, and 55.3% at 200–800 °C in flowing argon for Flu, Flu-PEG12K, and Flu-PEG30K, respectively (Fig. 8). The calculated PEG density of Flu-PEG12K and Flu-PEG30K was 0.031 and 0.032 PEG/nm^2, respectively. The PEG surface layer is often described as either a mushroom or brush-type layer. When the surface PEG density is relatively low, a mushroom-like structure is formed to maximize surface coverage [71]. As the PEG density increases, the PEG chains extend to avoid overlapping with other PEG molecules, resulting in a brush-type structure [71,72]. The distance between the PEGs of Flu-PEG30K was

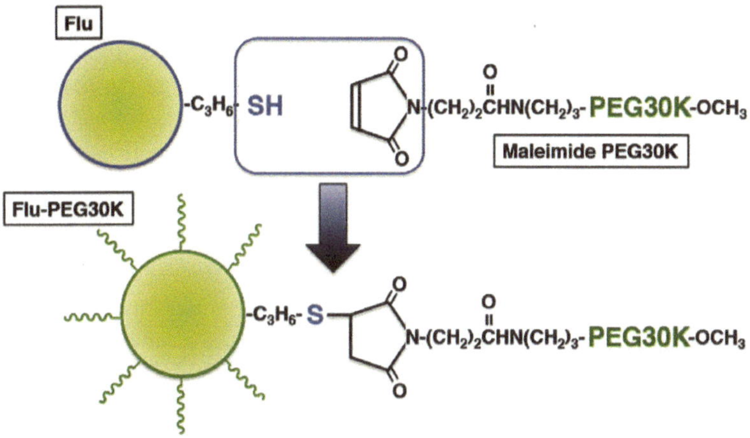

Fig. 7 The reaction scheme for the surface functionalization of Flu with methoxy-PEG-maleimide. *From M. Nakamura, K. Hayashi, M. Nakano, T. Kanadani, K. Miyamoto, T. Kori, K. Horikawa, Identification of polyethylene glycol-resistant macrophages on stealth imaging in vitro using fluorescent organosilica nanoparticles, ACS Nano 9 (2) (2015) 1058–1071.*

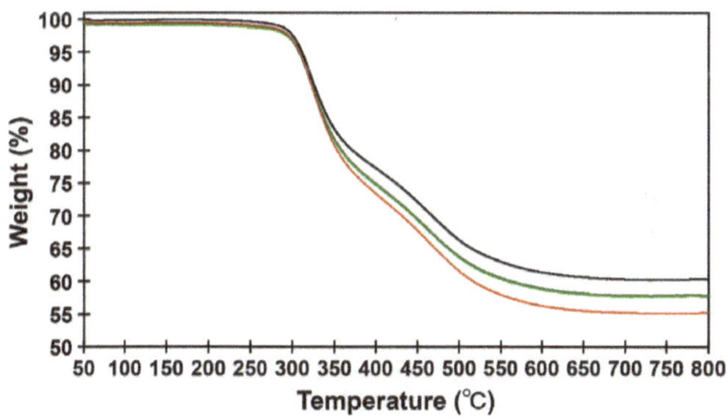

Fig. 8 Thermal gravimetric analysis of PEGylated nanoparticles. The data for Flu (black), Flu-PEG12K (green), and Flu-PEG30K (red) were obtained in flowing argon. *From M. Nakamura, K. Hayashi, M. Nakano, T. Kanadani, K. Miyamoto, T. Kori, K. Horikawa, Identification of polyethylene glycol-resistant macrophages on stealth imaging in vitro using fluorescent organosilica nanoparticles, ACS Nano 9 (2) (2015) 1058–1071.*

31.6 nm, smaller than the Flory radius of PEG30K (35 nm). These results indicated that Flu-PEG30K might adopt a brush-type structure, and had a more densely PEGylated surface than Flu-PEG12K.

In addition to biomolecules, surface functionalization of thiol-organosilica nanoparticles with nanomaterials is also reported recently [73–75].

Various types of hydrophobic ligand-capped nanoparticles were immobilized on the surface of thiol-organosilica nanoparticles directly. Functional nanoparticles such as iron oxide nanoparticles, quantum dots and Au nanoparticles were assembled on the surface of thiol-organosilica nanoparticles and then further overcoated with inorganosilicate layer to stabilize the assemblies and render them highly dispersible in water for medical imaging [74]. These studies demonstrated that surface of organosilica nanoparticles is also very useful for preparations of advanced multifunctionalized nanomaterials.

4. MEDICAL IMAGING OF ORGANOSILICA NANOPARTICLES

The ultimate goal of functionalization and biomedical application of nanoparticles is establishment of innovative medicine that produce cure of patients without any distress. The nanoparticles functionalized with imaging agents have demonstrated great potential to create innovative medical imaging. However, applications of nanoparticles to medicine are relative new challenges, and various basic subjects and questions are suggested. For instance, there are questions concerning nano-size effect of nanoparticles. The EPR effect is a unique feature that allows nanoparticles to preferentially accumulate and diffuse in tumor tissues [21–25]. However, there are various kinds of tumor cells and heterogeneity in tumor tissues. And the tumor pathology can change according to the stage of tumor progress. There are important questions which the size of nanoparticles can effective for tumor heterogeneity. The application of functionalized silica nanoparticle to medical imaging would address these kinds of questions.

4.1 Macroscopic and Mesoscopic Imaging With Multimodality

4.1.1 Macroscopic and Multimodal Imaging of Gastrointestinal Tract

Thiol-organosilica nanoparticles containing PI for multicolor fluorescence and X-ray computed tomography were applied for dual modal noninvasive functional gastrointestinal tract imaging [70]. Orally administered PI particles were observed in the gastrointestinal tract (GIT) using fluorescent imaging devices and X-ray CT. An in vivo fluorescence imaging system could detect the near-infrared fluorescence of PI particles in the GIT specifically. In addition, multipurpose zoom fluorescence microscopy was used to noninvasively visualize the real-time passage of particles and the movement of the GIT. The passage and distribution of particles over time in the GIT were demonstrated using X-ray CT. A correlation analysis between the fluorescent and X-ray CT data demonstrated the characteristics, limitations, and novel potential for noninvasive functional GIT dual modal imaging (Fig. 9).

4.1.2 Mesoscopic and Multimodal Imaging of Hepatic Tumor

We applied the thiol-organosilica nanoparticles containing iron oxide magnetic nanoparticles (MNPs) and rhodamine B, thiol-OS-MNP/Rho to multimodal imaging [64]. Injected mice revealed an alteration of the liver contrast in the MRI and a fluorescent pattern based on the liver histological structure at the level between macroscopic and microscopic fluorescent imaging (Fig. 10). We demonstrated a new approach to evaluate the liver at the macroscopic, microscopic level as well as the mesoscopic level using multimodal imaging. Thiol-OS-MNP/Rho were also applied to multimodal imaging of hepatic tumor of Long–Evans Cinnamon (LEC) rat. The MRI of LEC rats revealed tumors in the liver clearly and semiquantitatively due to a labeling of macrophages in liver. The fluorescent imaging showed abnormal fluorescent patterns of the liver at the mesoscopic level that was between macroscopic and microscopic level. Based on correlation analysis between optical imaging including fluorescent imaging and MRI, we found that the labeled macrophages located specific area in the tumor tissue and influenced the tumor size on MRI. In addition, histological observation showed the labeled macrophages related specific tissue in the pathological region (Fig. 11). We demonstrated a new approach to evaluate tumor tissue at the macroscopic and microscopic level as well as mesoscopic level using multimodal imaging providing new insight to tumor tissue evaluation.

4.2 Microscopic Imaging

A notable advantage of fluorescent imaging techniques is high temporal resolution, high sensitivity, multichannel imaging using multiple fluorescent probes, and relatively low-cost. Fluorescent silica nanoparticles are very promising for fluorescent microscopic imaging, including cellular, and tissue, because of their high fluorescence intensity and photostability. To investigate size- and surface-dependent effect of nanoparticles, organosilica nanoparticles would perform important contributions using their unique property. Because size control syntheses of organosilica particles are possible and their surface modifications are facile and effective. In addition, various kinds of fluorescent organosilica particles can be prepared, and enable highly sensitive and multichannel fluorescent imaging.

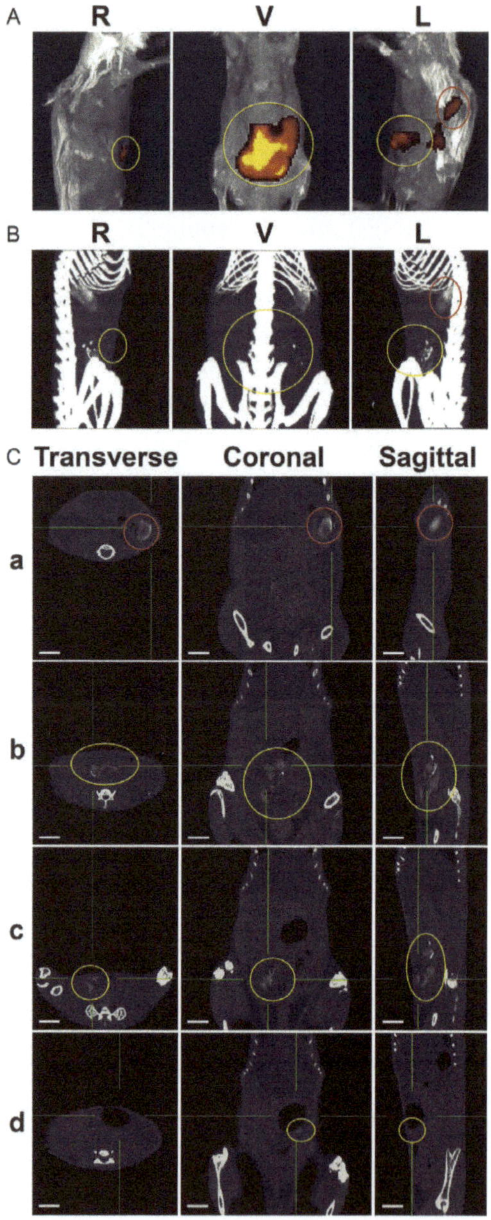

Fig. 9 Correlation analysis between in vivo fluorescence imaging and X-ray CT images. Fluorescence images (A) were compared with X-ray CT images of three-dimensions (B) and plain images (C) of oral administration of the PI particles. A left lateral view (L) of fluorescence imaging showed weak fluorescent intensity in the stomach ((A)L, red circles), but the CT images showed high CT signal intensity ((B)L and (C)a, red circles).

4.2.1 Nano-Biodistribution Imaging

We investigated size-dependent nano-biodistribution of fluorescent thiol-organosilica particles in mouse Peyer's patches (PPs) [61]. The PP is important lymphoid tissue for vaccination via oral administration. We prepared different sizes of fluorescent thiol-organosilica particles (95, 130, 200, 340, 695, and 1050 nm) to investigate size-dependent nano-biodistribution. We performed oral simultaneous dual particle administration using two kinds of fluorescent thiol-organosilica particles containing rhodamine B or fluorescein (Fig. 12). Quantitative analysis of dual administration studies demonstrated that the percent of fluorescence areas in PP subepithelial dome for 95, 130, 200, 340, 695, and 1050 nm as a function of 110 nm area was 124.0%, 89.1%, 73.8%, 20.2%, 9.2%, and 0.5%, respectively. Histological imaging and quantitative analysis revealed that thiol-organosilica particles uptake by PP were size-dependent. There was optimal size range for higher uptake of smaller particles (from 95 to 200 nm).

In addition, nano-biodistribution study can provide new knowledge in basic medicine. We have verified pathway through PP using fluorescent thiol-organosilica nanoparticles. In addition to the well-known transcellular pathway through M cell, high sensitive imaging using fluorescent thiol-organosilica nanoparticles could detect two novel pathways through mouse PP enterocytes, the transcellular pathway in PPs epithelium and the paracellular pathway in villi epithelium [61]. These pathways were detected and observed on fluorescent microscopy and electron microscopy (Fig. 13). Fluorescent thiol-organosilica nanoparticles have high fluorescent intensity, and can be detected even single nanoparticles on histological section. And silica nanoparticles can be observed on electron microscopic specimen effectively. These properties of silica nanoparticles would be useful for advanced nano-biodistribution study from micro- to nanoscale.

However, the X-ray CT images showed high CT signal intensity in left lower abdomen ((B)V and (C)d, green circles). All views of fluorescence imaging showed no fluorescence (6(B), green circles). The right lateral view (R) and ventral view (V) of fluorescence imaging showed strong fluorescent intensity in the right abdomen ((A)V and (A)R, yellow circles), but the CT images showed low CT signal intensity (maximum 843) ((B)L and 5(C)b, yellow circles). *From M. Nakamura, A. Awaad, K. Hayashi, K. Ochiai, K. Ishimura, Thiol-organosilica particles internally functionalized with propidium iodide as a multicolor fluorescence and x-ray computed tomography probe and application for non-invasive functional gastrointestinal tract imaging, Chem. Mater. 24 (2012) 3773–3779.*

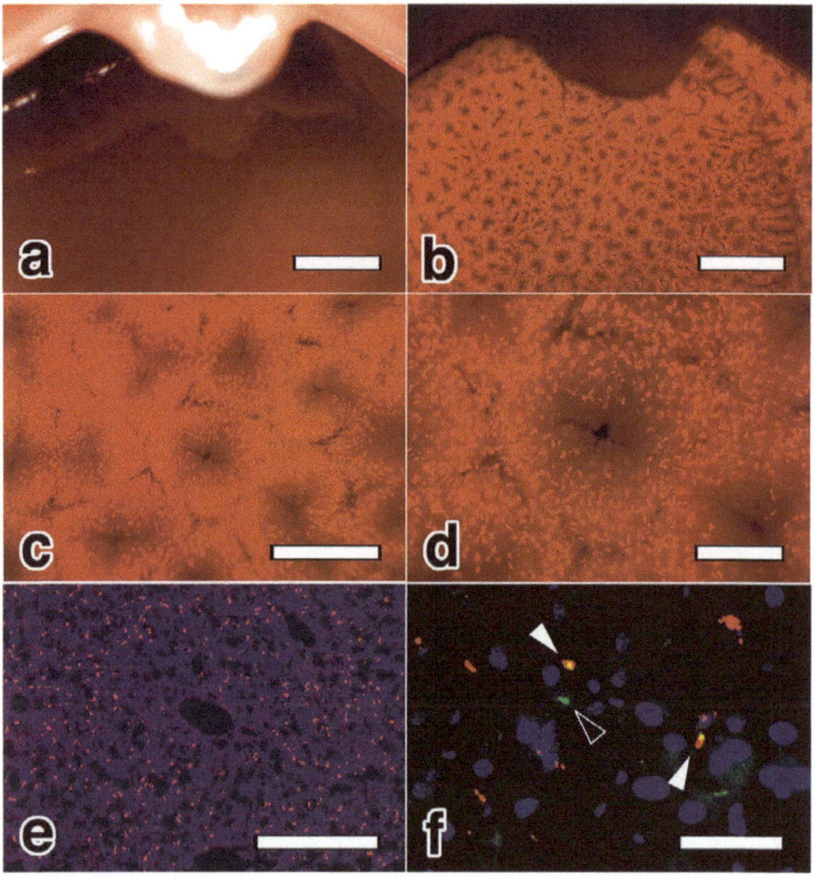

Fig. 10 Mesoscopic fluorescence imaging of normal mice. The mouse liver was observed by bright field (A) and fluorescent (B–D) with a multi-purpose zoom microscope. The liver showed fluorescence with a reticular pattern under low magnification (B). Under moderate and high magnification (C and D), the fluorescent labeled cells with thiol-OS-MNP/Rho were observed more in the peripheral region of the liver lobules than around the central veins (CV). The sections of mouse liver were observed by fluorescence microscopy (E). The labeled cells were distributed sparsely. In immunohistochemistry, with an anti-CD68 antibody (a marker associated with macrophage activation), some labeled cells with thiol-OS-MNP/Rho were stained with the antibody (white arrowhead). On the other hand, some CD68 positive cells did not show fluorescence from thiol-OS-MNP/Rho (F) (black arrowhead). Scale bars: (A, B) 2 mm, (C) 0.5 mm, (D, E) 0.2 mm, (F) 50 μm. *From M. Nakamura, K. Hayashi, H. Kubo, M. Harada, K. Izumi, Y. Tsuruo, T. Yogo, Mesoscopic multimodal imaging provides new insight to tumor tissue evaluation: an example of macrophage imaging of hepatic tumor using organosilica nanoparticles, Sci. Rep. 7 (1) (2017) 3953.*

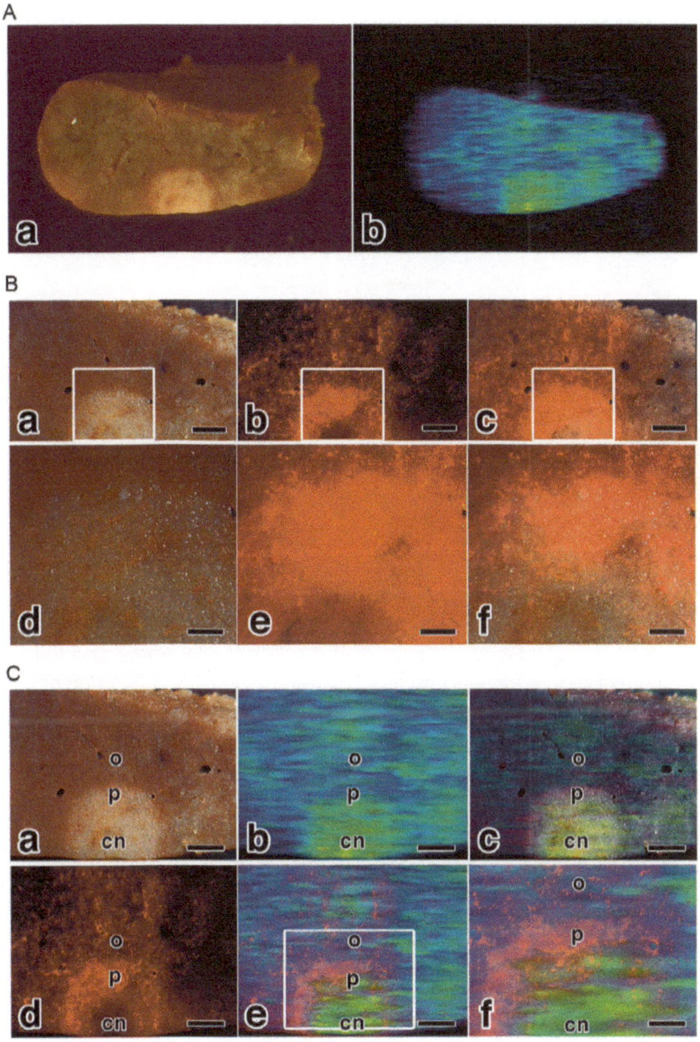

Fig. 11 Mesoscopic multimodal imaging of a sectioned liver of LEC rat. (A) The section of the pathologic liver showed a white nodule in the bright field image (a), and in the T2-weighted MRI (b). (B) The peripheral area (p) of the white nodule was observed on bright field (a and d) and on fluorescent imaging (b and e). The merged images (c and f) showed differences in the number of labeled cells among outside (o), peripheral (p), and central (cn) area of the white nodule. Scale bars: (a–c) 1 mm, (d–f) 0.25 mm. (C) Correlation analysis between the fluorescent imaging and MRI findings of the sectioned pathological liver. Compared with the bright field image (a), the findings of the MRI (b) and the merged image (c) showed that the size of the nodular area in the MRI was smaller than that of the bright field image. The merged images of the MRI and fluorescent imaging demonstrated the MRI signals were surrounded by fluorescent imaging signal (e and f). Scale bars: (a–e) 1 mm, (f) 0.5 mm. *From M. Nakamura, K. Hayashi, H. Kubo, M. Harada, K. Izumi, Y. Tsuruo, T. Yogo, Mesoscopic multimodal imaging provides new insight to tumor tissue evaluation: an example of macrophage imaging of hepatic tumor using organosilica nanoparticles, Sci. Rep. 7 (1) (2017) 3953.*

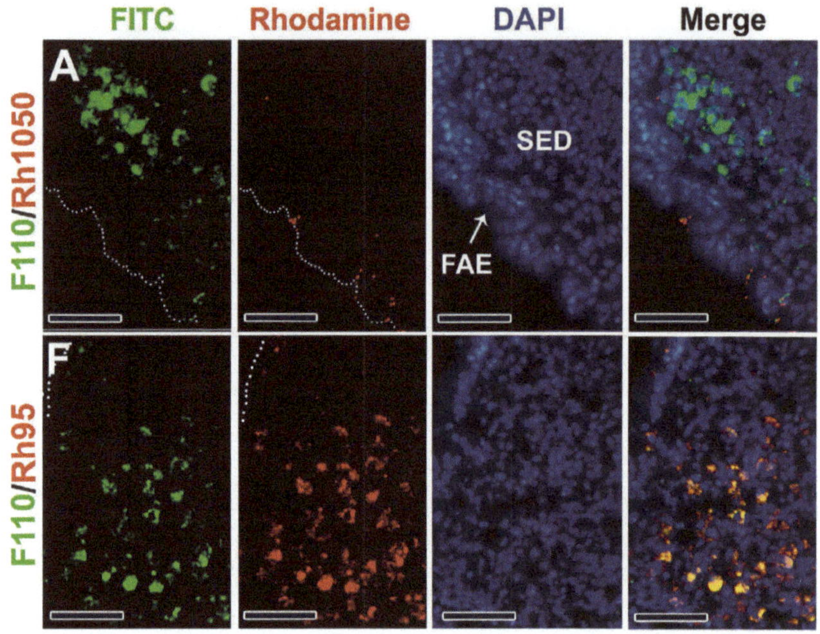

Fig. 12 Fluorescent microscope images of the subepithelial dome and follicle associated epithelium of mouse Peyer's patches 4 h after the oral administration of fluorescent thiol-organosilica particles in the simultaneous dual-particle administration study. The organosilica nanoparticles containing fluorescein (diameter 110 nm) (F110) was orally administered with various sizes of organosilica nanoparticles (95, 130, 200, 340, 695, and 1050 nm in diameter) containing rhodamine B (Rh). Fluorescence from F110 was observed in the subepithelial dome at much greater levels than those of Rh1050 (diameter 1050 nm). The fluorescence from Rh95 (diameter 95 nm) in the subepithelial dome was similar to and co-localized with those of F110 very well. Scale bar = 50 μm. *From M. Nakamura, K. Hayashi, H. Kubo, M. Harada, K. Izumi, Y. Tsuruo, T. Yogo, Mesoscopic multimodal imaging provides new insight to tumor tissue evaluation: an example of macrophage imaging of hepatic tumor using organosilica nanoparticles, Sci. Rep. 7 (1) (2017) 3953.*

4.2.2 Intracellular Nano-Localization Imaging

Surface dependent intracellular nano-localization of organosilica nanoparticles was also demonstrated using organosilica nanoparticles [41]. The nano-localization of epoxy-organosilica nanoparticles in peritoneal cells was examined by electron microscopy. Many epoxy-organosilica nanoparticles were contained in a large endosome in peritoneal cells. In addition, epoxy-organosilica nanoparticles located in the cytosol were observed. These findings indicate that some epoxy-organosilica nanoparticles escaped from the endoplasm to the cytosol. The surface charge can also be used to change the

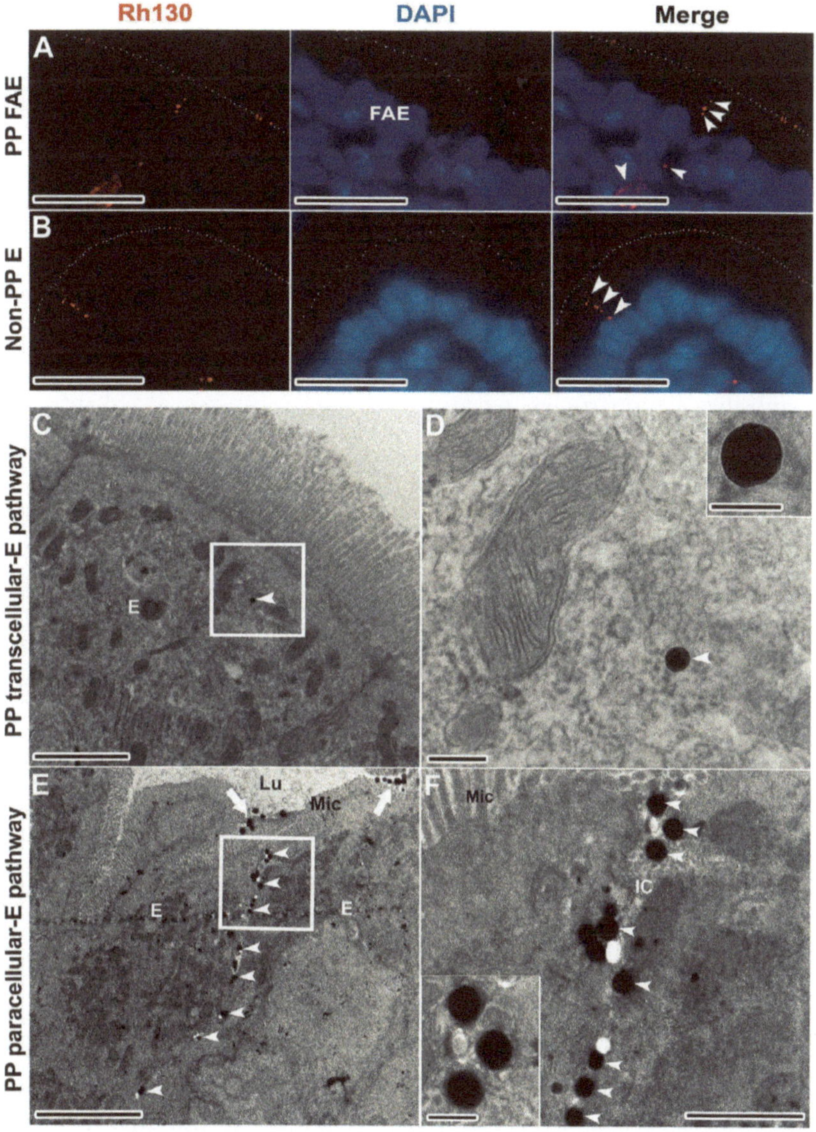

Fig. 13 Fluorescence microscopic and electron microscopic images of mouse Peyer's patches follicle associated epithelium (PP FAE) and non-Peyer's patches epithelium (Non-PP E) 4h after oral administration of fluorescent thiol-organosilica particles. (A) Linear florescence from particles (arrowheads) adjusting between two Peyer's patches enterocytes. (B) Linear florescence from particles (arrowheads) adjusting between two non-Peyer's patches enterocytes. (C and D) Three Peyer's patches enterocytes (E) and fluorescent thiol-organosilica particles (arrowheads) within the cytoplasm (white box (C) and enlarged (D)). (E and F) The particles in the lumen (Lu), between microvilli (Mic) (bold arrows), and adjusting in one line in the intercellular space (IC) between 2 enterocytes (E) (white box (E) and enlarged (F)). Scale bar = (A) and (B) 25 μm; (C) 2 μm; (D) 200 nm, enlarged 100 nm; (E) 5 μm; (F) 1 μm, enlarged 200 nm. *From M. Nakamura, K. Hayashi, H. Kubo, M. Harada, K. Izumi, Y. Tsuruo, T. Yogo, Mesoscopic multimodal imaging provides new insight to tumor tissue evaluation: an example of macrophage imaging of hepatic tumor using organosilica nanoparticles, Sci. Rep. 7 (1) (2017) 3953.*

intracellular localization: nanoparticles that are positively charged at endosomal pH (pH 4–5) escape from the endosomal compartment to the cytoplasm, whereas nanoparticles that are negatively charged at pH 4 are retained mostly in the endosomal compartment. The relationships between the surface properties of nanoparticles and cells are of great importance for the development of intracellular drug delivery systems. The surface charge of nanoparticles strongly affects the in vivo metabolism, biodistribution, and internalization as well as intracellular localization of the nanoparticles. Both epoxy-organosilica nanoparticles (positively charged) and thiol-organosilica nanoparticles (negatively charged), functionalized internally and on their surface, are expected to be useful for such applications.

4.2.3 Imaging of Molecular Trace

Dual fluorescent thiol-organosilica nanoparticles composed of quantum dots and rhodamine B in a thiol-organosilica layer were applied to imaging of molecular trace [62]. The dual fluorescent thiol-organosilica nanoparticles showed strong fluorescence, high photostability, and reduced photoblinking. Cells labeled with dual fluorescent thiol-organosilica nanoparticles were detected clearly in vivo. The nanoparticles surface-functionalized with anti-CD20 antibody could detect tumor antigen and showed unique movements on the cell surface of Raji cells (phenotype, $CD20^+/CD19^+/CD22^+$) (Fig. 14). The imaging of molecular trace would provide new aspect of cellular and molecular biology.

4.2.4 Functional Imaging of Macrophage

The interactions between macrophages and nanoparticles are very important for developments in nanomedicine because macrophages uptake the greater part of nanoparticles in vivo. Imaging system of interaction between macrophage and fluorescent thiol-organosilica nanoparticles can be a good model. The organosilica nanoparticles were applied to time-lapse fluorescence imaging to evaluate macrophage uptake function at the single cell level [60,70]. Time-lapse fluorescence imaging of mouse peritoneal macrophages using fluorescent thiol-organosilica nanoparticles containing rhodamine B clearly demonstrated cellular uptake [60]. The single cell quantitative analysis showed various patterns of uptake kinetics, indicating heterogeneous functionality of macrophages. Quantitative analysis of single endosomal uptake and movements in single cells could be also demonstrated.

Surface functionalized organosilica nanoparticles with PEG were also applied to single cell imaging of stealth function [70]. Because the uptake

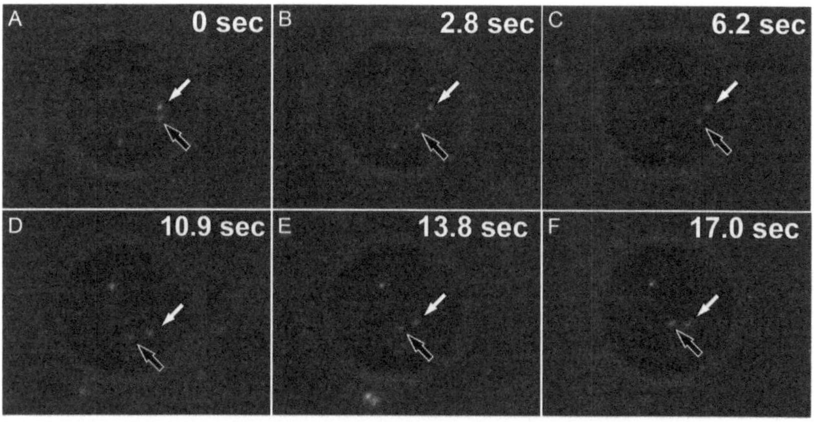

Fig. 14 Molecular trace imaging using dual fluorescent thiol-organosilica nanoparticles composed of quantum dots and rhodamine B surface functionalized with anti-CD20 antibody. The Raji cells were incubated with dual fluorescent thiol-OS-QDs functionalized with anti-CD20 antibody. Two dual fluorescent thiol-OS-QDs (white arrow and black arrow) showed different movements on the Raji cell. *From M. Nakamura, S. Ozaki, M. Abe, T. Matsumoto, K. Ishimura, One-pot synthesis and characterization of dual fluorescent thiol-organosilica nanoparticles as non-photoblinking quantum dots and their applications for biological imaging, J. Mater. Chem. 21 (2011) 4689–4695.*

level and kinetics of each macrophage were heterogeneous, we applied a simultaneous dual-particle administration study using two types of particles with different internal fluorescence and surface structure. For this imaging three types of fluorescent organosilica nanoparticles containing fluorescein (Flu), rhodamine B (Rho), or Rubpy (Rub) were prepared and evaluated. Transmission electron microscopy (TEM) revealed that these three nanoparticles had a spherical shape, narrow size distribution and good dispersions. The diameter and percentage of the CVs of Flu, Rho, and Rub were approximately 100 nm and 8%. Both bare Flu and Rho were taken up similarly (Fig. 15A). However, the uptakes of Flu-PEG30K and Rho into the cells (Fig. 15B) were different. Many cells showed uptake of only Rho, but some cells showed uptake of both Rho and Flu-PEG30K. Fig. 16 shows enlarged views of the pictures at 6 h of Fig. 15. As shown in Fig. 16A, the cells (A1 and A2) containing nanoparticles showed both Flu and Rho fluorescence. The macrophages incubated with Flu-PEG30K and Rho showed various patterns. Many cells showed only Rho fluorescence (B1 and B2), and some cells showed fluorescence of both Flu and Rho (B3–B6). Cells B3 and B4 had uptake of Rho and Flu-PEG30K at lower levels. Cells B5 and B6 showed similar uptake levels of Flu-PEG30K and Rho. Single-cell

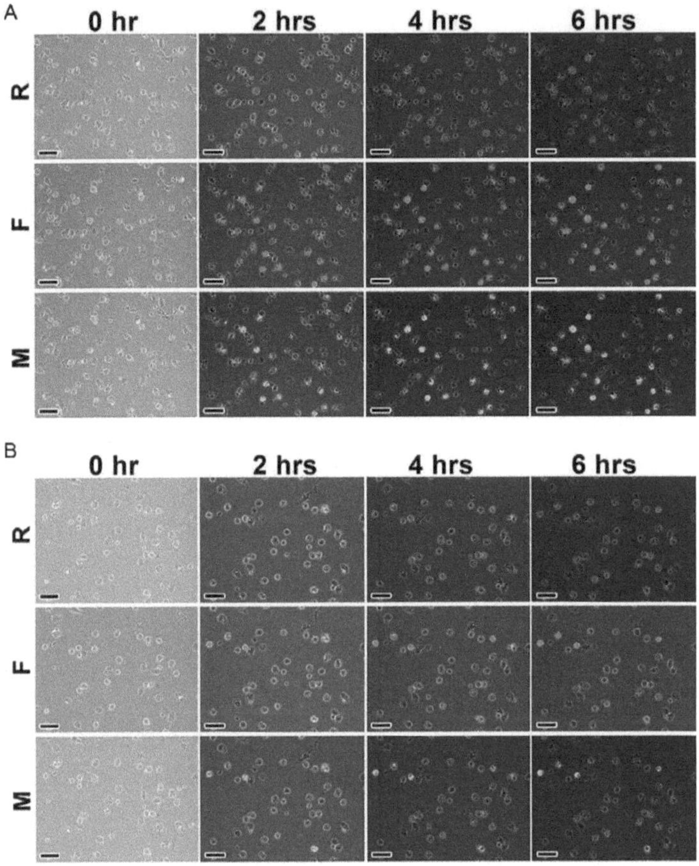

Fig. 15 Time-lapse imaging of the dual-particle administration study of macrophage uptake of fluorescent organosilica nanoparticles surface functionalized with PEG. The macrophages were incubated with a mixture of Flu and Rho (A) or with a mixture of Flu-PEG30K and Rho (B). Fluorescence images of Rub (R), Flu or Flu-PEG30K (F), and merged images (M) are shown from 0 to 6h. The scale bars represent 40 μm. *From M. Nakamura, K. Hayashi, M. Nakano, T. Kanadani, K. Miyamoto, T. Kori, K. Horikawa, Identification of polyethylene glycol-resistant macrophages on stealth imaging in vitro using fluorescent organosilica nanoparticles, ACS Nano 9 (2) (2015) 1058–1071.*

imaging and analysis revealed various patterns and kinetics of bare and PEGylated nanoparticle uptake. The PEGylated nanoparticles revealed a stealth function against most macrophages (PEG-sensitive macrophages); however, a stealth function against certain macrophages (PEG-insensitive macrophages) was not observed. We identified and characterized the PEG-resistant macrophages that could take up PEGylated nanoparticles at the same level as bare nanoparticles. We propose that PEG-insensitive

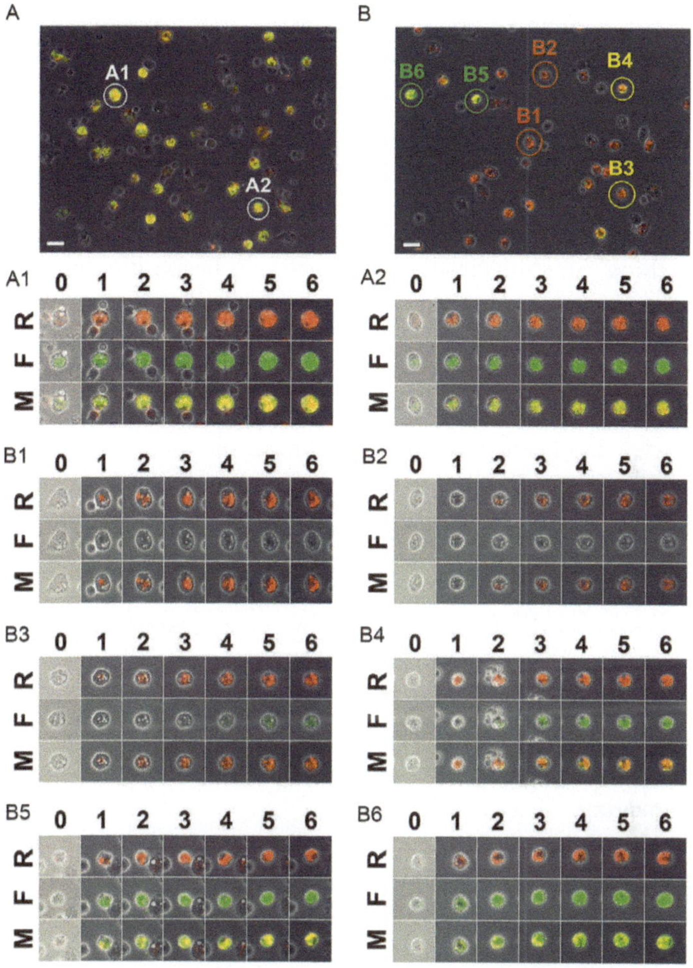

Fig. 16 Single-cell dual-particle images showing macrophage uptake of fluorescent organosilica nanoparticles surface functionalized with PEG. The macrophages were incubated with a mixture of Flu and Rho (A) or with a mixture of Flu-PEG30K and Rho (B). The fluorescence images of Rub (R), Flu or Flu-PEG30K (F), and merged images (M) of single cells are shown from 0 to 6 h. The scale bars represent 40 μm. Cells A1 and A2 showed similar uptake of Flu and Rho. Cells B1 and B2 showed uptake of Rho and almost no uptake of Flu-PEG30K. Cells B3 and B4 showed uptake of Rho and a lower uptake of Flu-PEG30K. Cells B5 and B6 showed a similar uptake of Flu, Rho, and Flu-PEG30K. *From M. Nakamura, K. Hayashi, M. Nakano, T. Kanadani, K. Miyamoto, T. Kori, K. Horikawa, Identification of polyethylene glycol-resistant macrophages on stealth imaging in vitro using fluorescent organosilica nanoparticles, ACS Nano 9 (2) (2015) 1058–1071.*

macrophages, including PEG-resistant macrophages, are a new concept and that this concept is important for understanding and regulating the immune responses against PEGylated nanoparticles.

5. OUTLOOK

According to a developmental outgrowth of nanomedicine, medical imaging using functionalized nanoparticles is becoming more positive, extensive and attractive. Advanced medical imaging such as multimodal imaging is being developed based on nanoparticles. In these trends, silica nanoparticles have unique properties and are pushed forward development for theranostics. Recently a novel type of silica nanoparticles, organosilica nanoparticles, has been developed. Common inorganosilica nanoparticles without organic residue as prepared are distinguished from organosilica nanoparticles. Both inorganosilica and organosilica nanoparticles have various advantages in biocompatibility as well as in synthesis. Size-controlled syntheses and multifunctionalization of organosilica nanoparticles are useful for advanced medical imaging. At present organosilica nanoparticles are used as imaging probes for basic nanomedicine such as multimodal imaging, nano-biodistribution, and functional imaging. An accumulation of knowledge of these basic nanomedicine is important and still required. Organosilica nanoparticles can be further multifunctionalized for advanced medical imaging.

Recently clinical developments of silica nanoparticles are one of the most active area in nanomedicine. An inorganosilica nanoparticle, Cornell dots (C dots), are applied for positron emission tomography (PET)–optical dual-modality imaging. C dots have received U.S. Food and Drug Administration (FDA) approval for a first-in-human clinical trial as a cancer-specific probe [76]. And clinical trial showed the particles are safe for human use [77]. Ultrasmall inorganosilica nanoparticles containing gadolinium have been developed for MRI-guided radiotherapy and also has been approved for clinical trial by FDA [78–80]. In addition, TheraSphere®, micro-sized inorganosilica particles loaded with yttrium-90, has been clinically developed. The FDA concluded the benefit of TheraSphere® in the treatment of hepatocellular carcinoma [81–84]. Actually silica particles are the most promising material for medical application. Clinical developments of organosilica nanoparticles also can be expected greatly. The approaches of internal and surface functionalization of organosilica nanoparticles with therapeutic reagents are simple and effective. Multifunctional abilities of organosilica nanoparticles have high potential for theranostics. Clinical applications of multifunctional organosilica nanoparticles with optimal multiplicity, nano-sized effect, and additive/multiplier effect would create novel medical imaging, and then theranostics.

ACKNOWLEDGMENTS

This work was supported partly by JSPS KAKENHI Grants (Nos. 17790356, 19510117, 21500409, 25350530, 16K01358), by a Grant for Practical Application of University R&D Results under the Matching Fund Method from the New Energy and Industrial Technology Development Organization (NEDO) of Japan, by a Research for Promoting Technological Seeds and Adaptable and Seamless Technology Transfer Program through target-driven R&D from JST, and by the JSPS Bilateral Programs.

REFERENCES

[1] S.M. Moghimi, A.C. Hunter, J.C. Murray, Nanomedicine: current status and future prospects, FASEB J. 19 (2005) 311–330.
[2] Y. Liu, H. Miyoshi, M. Nakamura, Nanomedicine for drug delivery and imaging: a promising avenue for cancer therapy and diagnosis using targeted functional nanoparticles, Int. J. Cancer 120 (2007) 2527–2537.
[3] M. Liong, J. Lu, M. Kovochich, T. Xia, S.G. Ruehm, A.E. Nel, F. Tamanoi, J.I. Zink, Multifunctional inorganic nanoparticles for imaging, targeting, and drug delivery, ACS Nano 2 (2008) 889–896.
[4] Y. Namiki, T. Fuchigami, N. Tada, R. Kawamura, S. Matsunuma, Y. Kitamoto, M. Nakagawa, Nanomedicine for cancer: lipid-based nanostructures for drug delivery and monitoring, Acc. Chem. Res. 44 (2011) 1080–1093.
[5] P. Yang, S. Gai, J. Lin, Functionalized mesoporous silica materials for controlled drug delivery, Chem. Soc. Rev. 41 (2012) 3679–3698.
[6] Z. Li, J.C. Barnes, A. Bosoy, J.F. Stoddart, J.I. Zink, Mesoporous silica nanoparticles in biomedical applications, Chem. Soc. Rev. 41 (2012) 2590–2605.
[7] E.C. Dreaden, A.M. Alkilany, X. Huang, C.J. Murphy, M.A. El-Sayed, The golden age: gold nanoparticles for biomedicine, Chem. Soc. Rev. 41 (2012) 2740–2779.
[8] T.L. Doane, C. Burda, The unique role of nanoparticles in nanomedicine: imaging, drug delivery and therapy, Chem. Soc. Rev. 41 (2012) 2885–2911.
[9] W.X. Mai, H. Meng, Mesoporous silica nanoparticles: a multifunctional nano therapeutic system, Integr. Biol. (Camb) 5 (1) (2013) 19–28.
[10] W.J. Mulder, F.A. Jaffer, Z.A. Fayad, M. Nahrendorf, Imaging and nanomedicine in inflammatory atherosclerosis, Sci. Transl. Med. 6 (239) (2014) 239sr1.
[11] N. Oliva, S. Unterman, Y. Zhang, J. Conde, H.S. Song, N. Artzi, Personalizing biomaterials for precision nanomedicine considering the local tissue microenvironment, Adv. Healthc. Mater. 4 (11) (2015) 1584–1599.
[12] P.P. Adiseshaiah, R.M. Crist, S.S. Hook, S.E. McNeil, Nanomedicine strategies to overcome the pathophysiological barriers of pancreatic cancer, Nat. Rev. Clin. Oncol. 13 (12) (2016) 750–765.
[13] X. Du, X. Li, L. Xiong, X. Zhang, F. Kleitz, S.Z. Qiao, Mesoporous silica nanoparticles with organo-bridged silsesquioxane framework as innovative platforms for bioimaging and therapeutic agent delivery, Biomaterials 91 (2016) 90–127.

[14] C. Kinnear, T.L. Moore, L. Rodriguez-Lorenzo, B. Rothen-Rutishauser, A. Petri-Fink, Form follows function: nanoparticle shape and its implications for nanomedicine, Chem. Rev. 117 (17) (2017) 11476–11521.
[15] H. Qiu, Y. Min, Z. Rodgers, L. Zhang, A.Z. Wang, Nanomedicine approaches to improve cancer immunotherapy, Wiley Interdiscip. Rev. Nanomed. Nanobiotechnol. 9 (5) (2017) e1456.
[16] J. Shi, P.W. Kantoff, R. Wooster, O.C. Farokhzad, Cancer nanomedicine: progress, challenges and opportunities, Nat. Rev. Cancer 17 (1) (2017) 20–37.
[17] F. Chen, G. Hableel, E.R. Zhao, J.V. Jokerst, Multifunctional nanomedicine with silica: role of silica in nanoparticles for theranostic, imaging, and drug monitoring, J. Colloid Interface Sci. 521 (2018) 261–279.
[18] H.E. Karahan, C. Wiraja, C. Xu, J. Wei, Y. Wang, L. Wang, F. Liu, Y. Chen, Graphene materials in antimicrobial nanomedicine: current status and future perspectives, Adv. Healthc. Mater. 7 (13) (2018) e1701406.
[19] V. Agrahari, V. Agrahari, Facilitating the translation of nanomedicines to a clinical product: challenges and opportunities, Drug Discov. Today 23 (5) (2018) 974–991.
[20] C.K.W. Chan, L. Zhang, C.K. Cheng, H. Yang, Y. Huang, X.Y. Tian, C.H.J. Choi, Recent advances in managing atherosclerosis via nanomedicine, Small 14 (4) (2018) 1702793.
[21] H. Maeda, Y. Matsumura, Tumoritropic and lymphotropic principles of macromolecular drugs, Crit. Rev. Ther. Drug Carrier Syst. 6 (1989) 193–210.
[22] J. Fang, H. Nakamura, H. Maeda, The EPR effect: unique features of tumor blood vessels for drug delivery, factors involved, and limitations and augmentation of the effect, Adv. Drug Deliv. Rev. 63 (2011) 136–151.
[23] H. Maeda, Toward a full understanding of the EPR effect in primary and metastatic tumors as well as issues related to its heterogeneity, Adv. Drug Deliv. Rev. 91 (2015) 3–6.
[24] H. Maeda, Polymer therapeutics and the EPR effect, J. Drug Target. 25 (9–10) (2017) 781–785.
[25] S.K. Golombek, J.N. May, B. Theek, L. Appold, N. Drude, F. Kiessling, T. Lammers, Tumor targeting via EPR: strategies to enhance patient responses, Adv. Drug Deliv. Rev. 130 (2018) 17–38.
[26] K. Van Dyck, R. Van Cauwenbergh, H. Robberecht, H. Deelstra, Bioavailability of silicon from food and food supplements, Fresenius J. Anal. Chem. 363 (1999) 541–544.
[27] R. Jugdaohsingh, Silicon and bone health, J. Nutr. Health Aging 11 (2007) 99–110.
[28] K.R. Martin, The chemistry of silica and its potential health benefits, J. Nutr. Health Aging 11 (2007) 94–97.
[29] S. Braley, The chemistry and properties of the medical-grade silicones, J. Macromol. Sci. Chem. 4 (1970) 529–544.
[30] A.K. Luhrs, W. Geurtsen, The application of silicon and silicates in dentistry: a review, Prog. Mol. Subcell. Biol. 47 (2009) 359–380.
[31] W. Stöber, A. Fink, E. Bohn, Controlled growth of monodisperse silica spheres in the micron size range, J. Colloid Interface Sci. 26 (1968) 62–69.
[32] M. Yanagi, Y. Asano, K. Kandori, K. Kon-no, Abstract: 39th Symposium on Division of Colloid and Interface Chemistry Chemical Society of Japan, Tokyo, 1986, 386.
[33] K. Osseo-Asare, F.J. Arriagada, Preparation of SiO_2 nanoparticles in a non-ionic reverse micellar system, Colloids Surf. 50 (1990) 321–339.
[34] R.P. Bagwe, C. Yang, L.R. Hilliard, W. Tan, Optimization of dye-doped silica nanoparticles prepared using a reverse microemulsion method, Langmuir 20 (2004) 8336–8342.
[35] S. Zanarini, E. Rampazzo, L. Della Ciana, M. Marcaccio, E. Marzocchi, M. Montalti, F. Paolucci, L. Prodi, Ru $(bpy)_3$ covalently doped silica nanoparticles as multicenter tunable structures for electrochemiluminescence amplification, J. Am. Chem. Soc. 131 (2009) 2260–2267.

[36] S.T. Selvan, T.T. Tan, D.K. Yi, N.R. Jana, Functional and multifunctional nanoparticles for bioimaging and biosensing, Langmuir 26 (2010) 11631–11641.
[37] M. Nakamura, K. Ishimura, Synthesis and characterization of organosilica nanoparticles prepared from 3-mercaptopropyltrimethoxysilane as the single silica source, J. Phys. Chem. C 111 (2007) 18892–18898.
[38] M. Nakamura, K. Ishimura, One-pot synthesis and characterization of three kinds of thiol-organosilica nanoparticles, Langmuir 24 (2008) 5099–5108.
[39] M. Nakamura, S. Ozaki, M. Abe, H. Doi, T. Matsumoto, K. Ishimura, One-pot synthesis, characterization, and biological applications of size-controlled thiol-organosilica nanoparticles prepared from 3-mercaptopropyltrimethoxysilane, Colloids Surf. B Biointerfaces 79 (2010) 19–26.
[40] T. Doura, F. Tamanoi, M. Nakamura, Miniaturization of thiol-organosilica nanoparticles induced by an anionic surfactant, J. Colloid Interface Sci. 526 (2018) 51–62.
[41] M. Nakamura, K. Ishimura, Size-controlled, one-pot synthesis, characterization, and biological applications of epoxy-organosilica particles possessing positive zeta potential as prepared, Langmuir 24 (2008) 12228–12234.
[42] M. Nakamura, A. Awaad, K. Hayashi, K. Ochiai, K. Ishimura, Thiol-organosilica particles internally functionalized with propidium iodide as a multicolor fluorescence and x-ray computed tomography probe and application for non-invasive functional gastrointestinal tract imaging, Chem. Mater. 24 (2012) 3773–3779.
[43] S. Santra, P. Zhang, K. Wang, R. Tapec, W. Tan, Conjugation of biomolecules with luminophore-doped silica nanoparticles for photostable biomarkers, Anal. Chem. 73 (2001) 4988–4993.
[44] X. Zhao, R.P. Bagwe, W. Tan, Development of organic-dye-doped silica nanoparticles in a reverse microemulsion, Adv. Mater. 16 (2004) 173–176.
[45] L.M. Rossi, L. Shi, F.H. Quina, Z. Rosenzweig, Stöber synthesis of monodispersed luminescent silica nanoparticles for bioanalytical assays, Langmuir 21 (2005) 4277–4280.
[46] Z. Ye, M. Tan, G. Wang, J. Yuan, Novel fluorescent europium chelate-doped silica nanoparticles: preparation, characterization and time-resolved fluorometric application, J. Mater. Chem. 14 (2004) 851–856.
[47] M. Tan, G. Wang, X. Hai, Z. Ye, J. Yuan, Development of functionalized fluorescent europium nanoparticles for biolabeling and time-resolved fluorometric applications, J. Mater. Chem. 14 (2004) 2896–2901.
[48] L. Xue, B. Li, Q. Fei, G. Feng, Y. Huan, Z. Shi, Carboxylate-modified squaraine dye doped silica fluorescent pH nanosensors, Nanotechnology 21 (2010) 215502.
[49] D. Zhang, Z. Wu, J. Xu, J. Liang, J. Li, W. Yang, Tuning the emission properties of Ru (phen)3 (2+) doped silica nanoparticles by changing the addition time of the dye during the Stöber process, Langmuir 26 (2010) 6657–6662.
[50] A. Yari, M. Saidikhah, Dye doped Eosin yellowish silica nanoparticles as novel fluorophore for a peroxyoxalate chemiluminescence system, J. Fluoresc. 22 (2012) 993–1002.
[51] Z.Z. Chen, L. Cai, X.M. Dong, H.W. Tang, D.W. Pang, Covalent conjugation of avidin with dye-doped silica nanoparticles and preparation of high density avidin nanoparticles as photostable bioprobes, Biosens. Bioelectron. 37 (2012) 75–81.
[52] A. Van Blaaderen, A. Vrij, Synthesis and characterization of colloidal dispersions of fluorescent, monodisperse silica spheres, Langmuir 8 (1992) 2921–2931.
[53] N.A.M. Verhaegh, A. van Blaaderen, Dispersions of rhodamine-labeled silica spheres: synthesis, characterization, and fluorescence confocal scanning laser microscopy, Langmuir 10 (1994) 1427–1438.
[54] A. Imhof, M. Megens, J.J. Engelberts, D.T.N. de Lang, R. Sprik, W.L. Vos, Spectroscopy of fluorescein (FITC) dyed colloidal silica spheres, J. Phys. Chem. B 103 (1999) 1408–1415.

[55] M. Nakamura, M. Shono, K. Ishimura, Synthesis, characterization, and biological applications of multifluorescent silica nanoparticles, Anal. Chem. 79 (2007) 6507–6514.
[56] X. He, J. Chen, K. Wang, D. Qin, W. Tan, Preparation of luminescent Cy5 doped core-shell SFNPs and its application as a near-infrared fluorescent marker, Talanta 72 (2007) 1519–1526.
[57] A. Liu, L. Wu, Z. He, J. Zhou, Development of highly fluorescent silica nanoparticles chemically doped with organic dye for sensitive DNA microarray detection, Anal. Bioanal. Chem. 401 (2011) 2003–2011.
[58] N. Zhang, E. Ding, X. Feng, Y. Xu, H. Cai, Synthesis, characterizations of dye-doped silica nanoparticles and their application in labeling cells, Colloids Surf. B Biointerfaces 89 (2012) 133–138.
[59] H. Ow, D.R. Larson, M. Srivastava, B.A. Baird, W.W. Webb, U. Wiesner, Bright and stable core-shell fluorescent silica nanoparticles, Nano Lett. 5 (2005) 113–117.
[60] M. Nakamura, K. Miyamoto, K. Hayashi, A. Awaad, M. Ochiai, K. Ishimura, Time-lapse fluorescence imaging and quantitative single cell and endosomal analysis of peritoneal macrophages using fluorescent organosilica nanoparticles, Nanomedicine 9 (2013) 274–283.
[61] A. Awaad, M. Nakamura, K. Ishimura, Imaging of size-dependent uptake and identification of novel pathways in mouse Peyer's patches using fluorescent organosilica particles, Nanomedicine 8 (2012) 627–636.
[62] M. Nakamura, S. Ozaki, M. Abe, T. Matsumoto, K. Ishimura, One-pot synthesis and characterization of dual fluorescent thiol-organosilica nanoparticles as non-photoblinking quantum dots and their applications for biological imaging, J. Mater. Chem. 21 (2011) 4689–4695.
[63] M. Nakamura, K. Hayashi, H. Kubo, T. Kanadani, M. Harada, T. Yogo, Relaxometric property of organosilica nanoparticles internally functionalized with iron oxide and fluorescent dye for multimodal imaging, J. Colloid Interface Sci. 492 (2017) 127–135.
[64] M. Nakamura, K. Hayashi, H. Kubo, M. Harada, K. Izumi, Y. Tsuruo, T. Yogo, Mesoscopic multimodal imaging provides new insight to tumor tissue evaluation: an example of macrophage imaging of hepatic tumor using organosilica nanoparticles, Sci. Rep. 7 (1) (2017) 3953.
[65] D.P. Jones, J.L. Carlson, P.S. Samiec, P. Sternberg, V.C. Mody, R.L. Reed, L.A.S. Brown, Glutathione measurement in human plasma: evaluation of sample collection, storage, and derivatization conditions for analysis of dansyl derivatives by HPLC, Clin. Chim. Acta 275 (1998) 175–184.
[66] M.E. Anderson, Glutathione: an overview of biosynthesis and modulation, Chem. Biol. Interact. 111–112 (1998) 1–14.
[67] S.S.M. Hassan, G.A. Rechnitz, Determination of glutathione and glutathione reductase with a silver sulfide membrane electrode, Anal. Chem. 54 (1982) 1972–1976.
[68] R. Hong, G. Han, J.M. Fernández, B.J. Kim, N.S. Forbes, V.M. Rotello, Glutathione-mediated delivery and release using monolayer protected nanoparticle carriers, J. Am. Chem. Soc. 128 (2006) 1078–1079.
[69] G.S. Irmukhametova, G.A. Mun, V.V. Khutoryanskiy, Thiolated mucoadhesive and PEGylated nonmucoadhesive organosilica nanoparticles from 3-mercaptopropyltrimethoxysilane, Langmuir 27 (2011) 9551–9556.
[70] M. Nakamura, K. Hayashi, M. Nakano, T. Kanadani, K. Miyamoto, T. Kori, K. Horikawa, Identification of polyethylene glycol-resistant macrophages on stealth imaging in vitro using fluorescent organosilica nanoparticles, ACS Nano 9 (2) (2015) 1058–1071.
[71] O. Tirosh, Y. Barenholz, J. Katzhendler, A. Priev, Hydration of polyethylene glycol-grafted liposomes, Biophys. J. 74 (1998) 1371–1379.

[72] R. Gref, M. Lück, P. Quellec, M. Marchand, E. Dellacherie, S. Harnisch, T. Blunk, R.H. Müller, 'Stealth' corona-core nanoparticles surface modified by polyethylene glycol (PEG): influences of the corona (PEG chain length and surface density) and of the core composition on phagocytic uptake and plasma protein adsorption, Colloids Surf. B Biointerfaces 18 (2000) 301–313.
[73] Z. Lu, J. Goebl, J. Ge, Y. Yin, Self-assembly and tunable plasmonic property of gold nanoparticles on mercapto-silica microspheres, J. Mater. Chem. 19 (2009) 4597–4602.
[74] Z. Lu, C. Gao, Q. Zhang, M. Chi, J.Y. Howe, Y. Yin, Direct assembly of hydrophobic nanoparticles to multifunctional structures, Nano Lett. 11 (2011) 3404–3412.
[75] S. Mühlig, A. Cunningham, S. Scheeler, C. Pacholski, T. Bürgi, C. Rockstuhl, F. Lederer, Self-assembled plasmonic core-shell clusters with an isotropic magnetic dipole response in the visible range, ACS Nano 5 (2011) 6586–6592.
[76] M. Benezra, O. Penate-Medina, P.B. Zanzonico, D. Schaer, H. Ow, A. Burns, E. DeStanchina, V. Longo, E. Herz, S. Iyer, J. Wolchok, S.M. arson, U. Wiesner, M.S. Bradbury, Multimodal silica nanoparticles are effective cancer-targeted probes in a model of human melanoma, J. Clin. Invest. 121 (7) (2011) 2768–2780.
[77] E. Phillips, O. Penate-Medina, P.B. Zanzonico, R.D. Carvajal, P. Mohan, Y. Ye, J. Humm, M. Gönen, H. Kalaigian, H. Schöder, H.W. Strauss, S.M. Larson, U. Wiesner, M.S. Bradbury, Clinical translation of an ultrasmall inorganic optical-PET imaging nanoparticle probe, Sci. Transl. Med. 6 (260) (2014) 260ra149.
[78] S. Kunjachan, A. Detappe, R. Kumar, T. Ireland, L. Cameron, D.E. Biancur, V. Motto-Ros, L. Sancey, S. Sridhar, G.M. Makrigiorgos, R.I. Berbeco, Nanoparticle mediated tumor vascular disruption: a novel strategy in radiation therapy, Nano Lett. 15 (11) (2015) 7488–7496.
[79] A. Detappe, S. Kunjachan, J. Rottmann, J. Robar, P. Tsiamas, H. Korideck, O. Tillement, R. Berbeco, AGuIX nanoparticles as a promising platform for image-guided radiation therapy, Cancer Nanotechnol. 6 (1) (2015) 4.
[80] S. Kotb, A. Detappe, F. Lux, F. Appaix, E.L. Barbier, V.L. Tran, M. Plissonneau, H. Gehan, F. Lefranc, C. Rodriguez-Lafrasse, C. Verry, R. Berbeco, O. Tillement, L. Sancey, Gadolinium-based nanoparticles and radiation therapy for multiple brain melanoma metastases: proof of concept before phase I trial, Theranostics 6 (3) (2016) 418–427.
[81] M.D. Liu, M.B. Uaje, M.S. Al-Ghazi, D. Fields, J. Herman, J.V. Kuo, N. Milne, T.H. Nguyen, N.S. Amsinghani, K.M. Tokita, F.Y. Tsai, D.J. Vajgrt, D.K. Imagawa, Use of Yttrium-90 TheraSphere for the treatment of unresectable hepatocellular carcinoma, Am. Surg. 70 (11) (2004) 947–953.
[82] R. Salem, R.J. Lewandowski, B. Atassi, S.C. Gordon, V.L. Gates, O. Barakat, Z. Sergie, C.Y. Wong, K.G. Thurston, Treatment of unresectable hepatocellular carcinoma with use of ^{90}Y microspheres (TheraSphere): safety, tumor response, and survival, J. Vasc. Interv. Radiol. 16 (12) (2005) 1627–1639.
[83] A.B. Benson 3rd, J.F. Geschwind, M.F. Mulcahy, W. Rilling, G. Siskin, G. Wiseman, J. Cunningham, B. Houghton, M. Ross, K. Memon, J. Andrews, C.J. Fleming, J. Herman, H. Nimeiri, R.J. Lewandowski, R. Salem, Radioembolisation for liver metastases: results from a prospective 151 patient multi-institutional phase II study, Eur. J. Cancer 49 (15) (2013) 3122–3130.
[84] E.W. Lee, L. Alanis, S.K. Cho, S. Saab, Yttrium-90 selective internal radiation therapy with glass microspheres for hepatocellular carcinoma: current and updated literature review, Korean J. Radiol. 17 (4) (2016) 472–488.